"I s...
sens...
sens...
trac...
even .... *sometimes experienced.* I can't say
enough kind words about the author or the book she
wrote."
Bella Four

"This is a reference book of additives so it can be
overwhelming at first. The organization is done very well
and makes it *easy to choose the worst of the worst.* It is
a book that helps you learn the terminolgy used on
packaging and lets you decide what you want to do with
the information. I didn't feel that it was the authors
opinions but rather documented information collected
through research and studies on record. ***You'll be
amazed at what our government allows in food when
Europe and other countries have outlawed these very
same additives.***"
Rebecca A. Johnson

***This is a practical handbook for the layperson and as
well, the health professional*** to use while shopping and
teaching others factual data of what is in the food items
they are purchasing. This book has a simple approach
and easy to carry size making it a good choice. Thank
you Christine for the tip on 'salted fish' being a
carcinogen. This is entirely true! I do recommend this
useful book for others.
Dr. H. Yellen, Escondido, CA

 "***What a shocker to see all the stuff that is put in our
food***! No wonder obesity is so rampant in the US!"
moshers4u

"*Changed my life!* About 1 year ago, I began having a lot of pain in my joints. After several doctors, I still did not have an explanation of why my joints hurt and why I was gaining weight even though I had always exercised and watched my diet. One doctor mentioned that red meat could trigger arthritis. I decided to find out more about nutrition and since I have a severe peanut allergy, I wanted to find out if any of the 'safe' foods I was eating could contain peanuts.

When I bought "Food Additives : A Shopper's Guide To What's Safe & What's Not," I was shocked to find out that many of the additives that I assumed were safe actually contained peanuts. *This book has become my "nutrition bible" I carry it with me wherever I go*. Since using this book, I have been able to avoid many additives that could be harmful to me. Within 1 month, I was able to begin an exercise program again without pain and I actually lost the weight I had gained.

Christine Hoza Farlow's book is *very easy to use* and small enough to fit in your pocket or purse. The material is straight-forward and lets you make quick choices that fit your nutritional needs whether you have allergies or simply want to avoid all the carcinogens that are popping up everywhere.

I have several other books about food additives, but I still seem to go back to this one. If you're looking for *a simple approach to understanding labels*, A Shopper's Guide To What's Safe & What's Not is definitely the book for you!"
Johanne Schwartz, Covina, CA

FOOD ADDITIVES

A Shopper's Guide To What's Safe & What's Not

Christine Hoza Farlow, D.C.

2013 Revised Edition

KISS For Health Publishing
Escondido, California
"Keep It Simple Secrets For Health"

FOOD ADDITIVES
A Shopper's Guide To
What's Safe & What's Not

Published by:
    KISS For Health Publishing
    P.O. Box 462335
    Econdido, CA 92046-2335

All rights reserved. No part of this book may be reproduced or transmitted in any form or by any means, electronic or mechanical, including photocopying, recording or by any information storage and retrieval system without written permission from the author, except for the inclusion of brief quotations in a review.

Copyright © 1993, 1997, 1999, 2001, 2004, 2007, 2013
by Christine Hoza Farlow, D.C.
All rights reserved.
Printed in the
United States of America

ISBN 978-0-9635635-8-3

Every effort has been made to insure the accuracy of the information presented in this book. However, nothing in this book should be construed as medical advice or used in place of medical consultation.

TABLE OF CONTENTS

How to Use This Book	7
Why You Should Use This Book	11
How the Safety Classifications Were Determined	15
IARC Classifications Regarding Cancer Risk	17
How to Read Labels	18
Genetically Modified Foods	21
Why Should You Care if You're Eating Genetically Modified Food?	24
Food Irradiation	25
Labeling of Fresh Produce	27
Food Additives	28
Other Food Additives	127
References	130
Order Form	140

HOW TO USE THIS BOOK

The codes below are to the left of each additive and indicate the safety of the additive.

* GRAS - Generally Recognized As Safe by the FDA. However, this *does not mean the additive is safe*. See page 12.

φ FDA approved colorant

S There is no known toxicity. The additive appears to be safe.

A The additive may cause allergic reactions.

C Caution is advised. The additive may be unsafe, poorly tested, or used in foods we eat too much of.

C1 Caution is advised for certain groups in the population, such as pregnant women, infants, persons with high blood pressure, kidney problems, etc.

X The additive is unsafe or very poorly tested.

The numbers below appear after some of the additives and indicate the purpose for which the additive is used, and the kinds of products in which you might find that additive.

1. <u>acidifier</u> - baked goods, candy, cheese, desserts, jelly, soft drinks.
2. <u>alkali</u> - baked goods, canned vegetables, chocolate, dairy products, frozen desserts, olives, tomato products.
3. <u>antibrowning agent</u> - fruits and vegetables.
4. <u>anticaking agent</u> - baking powder, nondairy creamer, salt, soft drink powders.
5. <u>antifoaming agent</u> - baked goods, beer, coffee creamer, fruit juices, jelly, dairy products, wine.
6. <u>antimycotic agent</u> - baked goods, cheese, dried fruit, jelly, processed meats, syrup.
7. <u>antioxidant</u> - bacon, baked goods, breakfast bars, butter, candy, canned fruits and vegetables, cream, fried foods, gelatin desserts, margarine, nuts, peanut butter, powdered soups, oil, salad dressing, shortening, spices, whipped toppings, foods containing artificial color or flavor.
8. <u>antistaling agent</u> - baked goods.
9. <u>artificial flavoring</u> - processed foods.
10. <u>artificial color</u> - baked goods, butter, cereal, candy, cheese, gelatin desserts, icing, maraschino cherries, margarine, meat, oranges, pasta, pudding, soft drinks.
11. <u>artificial sweetener</u> - processed foods.
12. <u>binder</u> - processed foods, snack foods.
13. <u>bleaching agent</u> - baked goods, cheese, fats, flour, oils.
14. <u>buffer</u> - baked goods, canned vegetables, cereals, cheese, chocolate, dessert mixes, jelly, ham, ice

cream, pasta, soft drinks, syrups.
15. <u>clarifying agent</u> - beer, soft drinks, vinegar, wine.
16. <u>color preservative</u> - processed foods.
17. <u>crystallization inhibitor</u> - oil and sugar products.
18. <u>dough conditioner</u> - baked goods.
19. <u>emulsifier</u> - baked goods, cake mixes, candy, chocolate, dairy products, ice cream, margarine, nondairy creamer, peanut butter, pickles, processed meats, shortening, toppings.
20. <u>enzymes</u> - cheese, fats, hydrolyzed protein.
21. <u>fat substitute</u> - dairy-type products, fried foods, frozen desserts, low fat products, margarine, mayonnaise.
22. <u>filler</u> - processed foods.
23. <u>firming agent</u> - fruits and vegetables.
24. <u>flavor carrier</u> - candy, soft drinks, syrups.
25. <u>flavor & color solvent</u> - coffee decaffeination, herbs, spices.
26. <u>flavor enhancer</u> - canned vegetables, fruit drinks, gelatin desserts, gravy, ice cream, jelly, meat products, nondairy creamer, sauces, soft drinks, soups, soup mixes.
27. <u>flavoring agent</u> - processed foods.
28. <u>foaming agent</u> - frozen desserts.
29. <u>food coating/glaze</u> - candy, on fruits and vegetables.
30. <u>food grade shellac</u> - candy.
31. <u>fungus source of enzymes</u> - baked goods, beef.
32. <u>humectant</u> - baked goods, candy, diet food, ice cream, jelly, shredded coconut, soft drinks.
33. <u>leavening</u> - baked goods, brewed drinks, cake mixes, flour.
34. <u>maturing agent</u> - baked goods.
35. <u>meat tenderizer</u>

36. milk protein
37. milk sugar
38. natural food color - see artificial color.
39. neutralizer - dairy products, processed foods.
40. preservative - processed foods.
41. propellant gas - whipping cream, vegetable sprays.
42. protein extender - processed foods.
43. salt substitute
44. sequestrant - beverages, fats, oils.
45. softener - processed foods.
46. stabilizer - baked goods, cocoa, fruit drinks, ice cream, pudding.
47. stimulant - chocolate, cocoa, coffee, soft drinks.
48. suspending agent - processed foods.
49. sweetener - processed foods.
50. texturizer - canned goods, frozen desserts, ice cream, other processed foods.
51. thickener - baby food and formula, ice cream, jelly, pudding, salad dressing, soft drinks, soup, yogurt.
52. yeast food - baked goods, beer, wine; may contain free glutamates.

WHY YOU SHOULD USE THIS BOOK

This book will guide you when you're shopping. Carry it in your pocket or purse so you can make informed choices about the foods you buy.

There are more than 3000 different chemicals purposefully added to our food. Safety testing of these chemical additives is generally done by the company that wants to produce the chemicals or use them in the foods they produce.

The Delaney Clause of the 1958 Food Additives Amendment states that any additives shown to cause cancer in humans or animals are not permitted in our food. However, because of political pressure the FDA allows small amounts of cancer causing substances to be used in foods.

Even if all food additives used in our foods were safe individually, rarely does any food have only one additive in it. Testing for additive safety has been done for single additives, not combinations. Additives safe individually may be harmful in certain combinations. ***Nobody knows the effects of the many different additives used in the thousands of different combinations, the results that will occur from eating numerous different products together, with different chemical additives, or the consequences of consuming these ingredients or products over time.***

This book lists over 1000 of the most common food additives. Each additive is preceded by a code representing its safety and the advisability of its use. The code indicates if the additive is <u>G</u>enerally

Recognized As Safe (GRAS) by the FDA, an evaluation of its safety and advisability of its use, *independent of the FDA evaluation*. The codes are listed on page 7 under **How To Use This Book**.

The GRAS classification of safety by the FDA does not guarantee that the additive is safe.

In the past, the FDA evaluated food additives to determine if they qualified for GRAS status. They were evaluated based upon their ability to cause cancer and harmful reproductive effects, often ignoring other harmful outcomes. Consequently, because of the incomplete evaluation of the potential adverse effects of the additives determined to be GRAS, the GRAS classification did not and still does not absolutely guarantee that an additive is actually safe.

In 1997, the FDA stopped reviewing additives for GRAS status. The agency proposed to abolish the FDA review for GRAS status and replace it with the GRAS Notification Program whereby the manufacturer would inform the FDA of the GRAS status of the additive. Although a final ruling has not yet been issued on this procedure, this is how the GRAS status has been determined since the proposal.

So now, manufacturers of the additives are determining the GRAS status of the additives they produce or wish to use in products they produce. Manufacturers have been known to be less than honest in their testing and reporting to the FDA of the truth about the safety of ingredients they wish to have approved. Therefore, additives that have received GRAS status since this new notification procedure has been operating may be less likely to even meet up to the weak standards that the FDA followed when it was doing the reviews itself.

In addition, a number of formerly GRAS additives have been removed from the GRAS list *after they were found to be harmful*. This was prior to the adoption of the notification system for determining GRAS status. It is highly likely that some additives in common use now, and considered safe, would be found harmful if they were thoroughly tested by an independent laboratory. However, this is unlikely to happen with the manufacturers determining their own GRAS status.

GRAS is not a classification you can rely on for determining if the additives you use are safe. It was included in this book so that you know what GRAS means and more importantly so that you know that you CANNOT rely on it for true safety information about food additives.

Therefore, you will find that **not all GRAS additives are rated as safe (S)** in the food additives listing because the research, or lack of it, does not support a safe rating. This is explained in more detail below in **How The Safety Classifications Were Determined**. Also, for more information, see the **References** at the end of the book.

Some of the additives are followed by a number or series of numbers. These numbers represent a code for the uses of the additives. The codes are listed on pages 8 - 10.

The additive listing may also include some of the adverse effects that may be associated with its consumption, and if the additive has not been adequately tested.

HOW THE SAFETY CLASSIFICATIONS WERE DETERMINED

Many sources were used to determine ingredient safety classifications. Often, they were not in complete agreement. Those indicating the most severe reactions were given the most weight.

Ingredients are rated **X** if:
- listed as carcinogens by the EPA, NTP, NIOSH, Carcinogenic Potency Project
- the International Agency for Research on Cancer (IARC) gives them a Group 1, 2A or 2B rating
- on EPA Hazardous Substances list
- not carcinogenic, but may form a carcinogen by reacting with another ingredient in the product
- not carcinogenic, but may be contaminated with a carcinogen in the production of the ingredient
- known to be unsafe for various reasons
- there's no safety data

Ingredients are rated **C** if:
- IARC gives them a Group 3 rating
- they may cause a variety of mild to moderate adverse effects
- they're considered safe, but there's inadequate safety data available

Ingredients are rated **C1** if:
- they may be harmful for certain groups of the population, i.e., children or pregnant women.

Ingredients are rated **S** if:
- known to be safe, supported by safety data
- known safe for the general population, but some people may have a mild reaction to the ingredient.

The safest products are products with the fewest ingredients and with ingredients rated S. Remember, even if all of the ingredients are safe individually, rarely does any product contain only one ingredient. Safety testing has only been done for individual ingredients, not combinations of ingredients. Ingredients safe individually may be harmful in certain combinations. *Nobody knows the effects of the many different ingredients used in the thousands of different combinations, the effects of eating numerous different products together, with different chemical additives, or the cumulative effects of consuming these ingredients or products over time.*

Not All Safety Ratings Agree With the FDA

Numerous additives that are Generally Recognized As Safe (GRAS) by the FDA are **NOT** classified as safe. *Many of the GRAS additives were never tested for safety*. The safety data obtained by researching these ingredients did not support a safe rating. *GRAS does not mean the ingredient is safe.*

IARC CLASSIFICATIONS REGARDING CANCER-CAUSING RISK

The International Agency for Research on Cancer (IARC) evaluates data from scientific studies to determine if there is a risk that the chemicals or mixtures are carcinogenic and classifies them into the following categories:

Group 1 – human carcinogen
Group 2A – probable human carcinogen
Group 2B – possible human carcinogen
Group 3 – cannot be classified as a human carcinogen
Group 4 –probably not a human carcinogen

For more detailed information and explanations of the IARC Group classifications, see:

http://monographs.iarc.fr/ENG/Preamble/currentb6e valrationale0706.php

Ingredients listed in this book are classified as a carcinogen only if there is supportive evidence of its carcinogenic status from IARC, or other agencies that are qualified to determine carcinogenic status. The IARC Group classification is listed for individual ingredients in the **Food Additives** listing.

HOW TO READ LABELS

Finding the ingredients on the label and being able to read them can be a challenge. They are often hidden under a flap of packaging material in very tiny print, barely readable without a magnifying glass.

Often the package has statements like "NATURAL FRUIT FLAVORS, with Real Fruit Juice," or ALL NATURAL INGREDIENTS and NO PRESERVATIVES ADDED. This **DOES NOT** mean there are no harmful additives in the product. The manufacturer hopes you'll think these are healthy, natural products, but if you read the list of ingredients and compare each additive with the additives listed in this book, you'll see it's not true.

Ingredients are listed on the label in order of predominance by weight ... the ingredient that weighs the most is listed first, the ingredient that weighs the least is listed last.

Here's a general rule of thumb: if the list of ingredients is long, there's probably a lot of chemical additives in the product, and you're risking your health by eating it. If the list of ingredients is short, it may or may not have harmful additives in it, so read the ingredients carefully before you decide to purchase the product.

Nutrition Facts gives you information on calories, and grams of fat, cholesterol, sodium, carbohydrates, dietary fiber, sugars, and protein in each serving. Here's some useful conversions to help you use and understand this information better.

1 gram of fat = approximately 9 calories.
1 gram of protein = approximately 4 calories.
1 gram of carbohydrate = approximately 4 calories.
4 grams of sugar = 1 teaspoon of sugar.

Let's take an example of a 123 calorie snack with 7 grams of fat, 2 grams of protein and 13 grams of carbohydrate of which 12 grams of the carbohydrate is sugar.

To get fat calories, multiply 9x7=63 calories from fat.

To get percentage of fat, divide 63 fat calories by 123 snack calories to get 51% fat.

Use the same procedure for protein and carbohydrate, using 4 calories per gram instead of 9.

To get the number of teaspoons of sugar in the snack, divide 12 grams of sugar in the snack by 4, to get 3 teaspoons of sugar.

$$\frac{12 \text{ grams of sugar}}{4 \text{ grams per teaspoon}} = 3 \text{ teaspoons of sugar}$$

The example is summarized below.

	grams	calories	Percent
Fat	7	63	51%
Protein	2	8	7%
Carbohydrate (total)	13	52	42%
sugar	12	48	39%
other	1	4	3%

Buying a packaged product in a health food store does not guarantee that it will be free of harmful additives. The only way to be sure there are no harmful additives in the food you buy is to read every label of every package and buy fresh, whole organic foods whenever possible.

GENETICALLY MODIFIED (GM) FOODS

If you're eating conventionally grown food, non-organic, you're probably eating some genetically modified food without even knowing it. The first genetically modified foods hit the grocery stores in 1994.

In the U.S., about 70% of the foods on the grocery store shelves contain genetically modified ingredients. More than 80% of cheese is produced using a genetically engineered enzyme, chymosin. Much of the meat, poultry, eggs and dairy products are from animals fed genetically modified feed, and 80-90% of the milk is mixed with milk from cows injected with growth hormone, rbGH. Honey and bee pollen may come from genetically modified sources of pollen.

"Some of the foods that may contain GM ingredients include infant formula, salad dressing, bread, cereal, hamburgers and hotdogs, margarine, mayonnaise, crackers, cookies, chocolate, candy, fried food, chips, veggie burgers, meat substitutes, ice cream, frozen yogurt, tofu, tamari, soy sauce, soy cheese, tomato sauce, protein powder, baking powder, alcohol, vanilla, powdered sugar, peanut butter, enriched flour and pasta." Also confectioner's glaze, vanilla extract, malt and white vinegar.

The foods most commonly genetically modified are soy (94%), cotton from which cottonseed oil is made (90%), canola (90%), sugarbeets (95%), corn (88%), Hawaiian papaya (more than 50%), over 24,000 acres of zucchini and yellow squash, and some alfalfa, and Quest® brand tobacco.

If approved, GMO salmon could be in stores by 2014.

Products that are not organic, and contain ingredients like sugar, molasses, soy oil, soy flour, soy lecithin, soy protein, soy protein concentrate, soy protein isolate, soy protein supplements, soy isoflavones, textured vegetable protein (TVP), cottonseed oil, corn oil, corn flour, corn gluten, corn masa, corn starch, corn meal, corn syrup, corn sugar, high fructose corn syrup, fructose, canola oil, vegetable oil, vegetable shortening, vegetable fat, margarine, maltodextrin, dextrose, lactic acid, citric acid are likely to be derived from genetically modified sugar beets, soy, cotton, corn or canola.

At this time, genetically modified ingredients in packaged foods are not labeled in the U.S., so there is no way of knowing from the label, if the ingredients in packaged or canned foods have been genetically modified, unless they're organic. Organically grown foods cannot be genetically modified.

The Institute for Responsible Technology has a list of foods by brand that contain genetically modified ingredients and those that do not at http://www.nongmoshoppingguide.com

The Non-GMO Project is a third party group that verifies non-GMO claims by food producers. Participating companies can display the Non-GMO Verified label on their products.

Although there is great consumer demand for labeling of GE ingredients in the U.S., GM foods and ingredients are still unlabeled. Recent attemps to to pass GMO labeling laws in several states died when the states were threatened with lawsuits by Monsanto. In November 2012, the California ballot initiative to require the labeling of GMOs in our food, was narrowly defeated when Monsanto, and conventional food companies outspent the pro-labeling camp 6 to 1 and flooded the media with misinformation. This has increased awareness of GMOs and has triggered a movement in numerous other states to require their labeling.

A list of companies who oppose GMO labeling and those that support it can be found at http://healthyeatingadvisor.com/blog/?p=1749 and http://www.anh-usa.org/boycott-companies-fighting-prop-37/

It's time to vote with your dollars. Boycott the companies who contributed to defeat GMO labeling, including the organic brands owned by major conventional food companies. Buy from the companies who supported labeling of GMOs and your right to know what's in your food. Currently, approximately 50 countries around the world require the labeling of GMOs in food products, but not the U.S. or Canada.

WHY SHOULD YOU CARE IF YOU'RE EATING GM FOOD?

The first ever long term study on the safety of eating GM corn was published in November 2012. The results showed kidney and liver damage and breast tumors.

Regulation is so lax in the genetic engineering sector, that some of the most basic and simple questions about GM foods haven't even been asked. Government regulations favor the biotech industry, not the health of the consumer.

Neither the FDA nor the industry wants to take responsibility for the safety of GM foods.

GM foods do not require FDA approval before they're allowed on the market. The FDA puts responsibility for safety on the company developing the food.

Phil Angell, Monsanto's director of corporate communications, in a conversation with Michael Pollan, author of the article "Playing God in the Garden," said that "Monsanto should not have to vouch for the safety of biotech food. Our interest is in selling as much of it as possible. Assuring its safety is the FDA's job."

Jeffrey Smith, in his book, *Genetic Roulette*, documents reports of hundreds of people with toxic or allergic reactions, thousands of sick, sterile or dead animals, and countless ways in which GM foods are inherently dangerous and virtually untested.

FOOD IRRADIATION

Food irradiation is bombarding food with ionizing radiation to kill bacteria. It destroys vitamins and minerals and kills all living cells in the irradiated food.

During the irradiation process, chemical reactions occur causing the formation of harmful chemicals, like formaldehyde and benzene, both of which cause cancer.

Irradiated food is not radioactive. It becomes what is scientifically called radiometric. Eating irradiated foods "can have effects that mimic those of actual exposure to ionizing radiation." So even though the food is not radioactive, eating it gives you an indirect exposure to radiation.

According to FDA policy, irradiated foods for retail sale are required to be labeled with the **radura**, or with the words "treated with ionizing radiation." However, the FDA is proposing eliminating the labeling of irradiated food if it does not cause material change in the food. This would lead to an increase of irradiated foods on the market.

Foods currently approved for irradiation include beef, lamb and pork, poultry, lettuce and spinach, oysters, mussels, scallops and clams, fresh fruits and vegetables, dry or dehydrated spices and seasonings, fresh shell eggs, seeds for sprouting.

The irradiated products actually on the market include herbs, herb teas, garlic and spices, some ingredients for nutritional supplements, and some ground beef and poultry. Various fruits and vegetables from Florida, tropical fruits and Hawaiian papayas are also irradiated.

As of 2004, irradiated meat is allowed in school lunch programs, without notifying the parents. However, as of February 2009 it does not appear that there are many if any schools using irradiated meat.

Imports of irradiated produce have been permitted in the U.S. since 2003. Labeling of these foods is required, but the U.S. inspects only about 2% of imported food. So, it's possible that unlabeled irradiated food is entering the country.

On May 11, 2012, the FDA denied a request for a review of the use of irradiation on sprouting seeds.

No long term studies to prove the safety of food irradiation have been done. The long-term health effects from eating irradiated food are unknown.

Other names for the process of food irradiation are electronic pasteurization, cold pasteurization or pasteurization with x-rays.

You can avoid irradiated foods by buying organically grown food.

LABELING OF FRESH PRODUCE

Have you ever wondered what those pesky little stickers mean that are attached the fruits and vegetables you buy?

According to the International Federation for Produce Standards, the numbers on the stickers are Price Look Up (PLU) codes and are a voluntary option intended for the benefit of the cashiers and those responsible for the store inventory.

However, you can benefit from these codes too. Learn to read the stickers and you can tell if the fresh fruits and vegetables you buy are commercially grown or organically grown by the numbers on the stickers.

If the item doesn't have a sticker on it, just look at the sign for a 4- or 5-digit number.

If the number is:
- 4 digits, it's conventionally grown.
- 5 digits starting with **9**, it's organically grown.

This is an easy way to distinguish between the conventionally grown and organic produce.

There's also a 5 digit code starting with **8** for genetically modified. However, you probably won't find any sticker numbers beginning with an 8 since the industry is so opposed to labeling GMO foods.

FOOD ADDITIVES

* C A   <u>Acacia gum</u> – 17, 19, 46; may cause skin rashes; not adequately tested.

S   <u>Acai Berry Extract</u> – 7; antioxidant superfood.

X A   <u>Accent</u> – see MSG.

X   <u>Acesulfame-K</u> – 11, "Sunette"; may cause low blood sugar attacks, diarrhea, cramps, nausea, vomiting; may cause kidney, liver damage, thalassemia, hemachromatosis, sickle cell anemmia; has caused cancer in lab animals; not tested on humans.

X   <u>Acesulfame-potassium</u> – same as acesulfame-K.

X A   <u>Acetal</u> –9, 27; may cause breathing difficulty, heart problems, high blood pressure; central nervous system depressant.

* X   <u>Acetaldehyde</u> – 9, 27; irritant to mucous membranes, central nervous system depressant, large doses may cause death; possible carcinogen, IARC Group 2B.

X   <u>Acetamide</u> – 27; skin, eye, respiratory irritant; may cause stomach irritation in large amounts; IARC Group 2B

C   <u>Acetanisole</u> – eye, skin, respiratory, gastrointestinal irritant; may be harmful if swallowed; not adequately tested.

* C   <u>Acetic acid</u> – 1, 27; may cause gastro-intestinal distress, skin rashes, eye irritation.

* C   <u>Acetoin</u> – 27; eye, skin, respiratory irritant; may be harmful if swallowed or absorbed through skin; not adequately tested.

28

C	<u>Acetone</u> – skin, eye irritant; large amounts ingested may cause nausea, vomiting, abdominal pain; small amounts unlikely to produce harmful effects.
C	<u>Acetophenone</u> – 27; eye, skin respiratory irritant; may cause nausea, abdominal pain, dizziness, headaches, loss of coordination, central nervous system harm; derived from coal tar.
* C	<u>Acetyl methylcarbinol</u> – 27; see acetoin.
C A	<u>Acetylated mono- and diglycerides</u> – 19, see mono- & diglycerides.
C	<u>Acetylpyrazine</u> – 27; eye, skin, respiratory, digestive irritant; not adequately tested.
C	<u>Acidified sodium chlorite</u> – anti-microbial in water applied to processed fruits and vegetables; not adequately tested.
X	<u>Aclame</u> – see alitame.
* S	<u>Aconitic acid</u> – 27.
X	<u>Acrylamide</u> – possible carcinogen, IARC Group 2A.
S	<u>Activated Carbon</u> – used in processing foods.
* C	<u>Adipic acid</u> – 1, 27, 40; skin, eye, digestive and respiratory irritant; may cause autonomic nervous system disorders.
C	<u>Adonitol</u> – see sugar alcohols.
X	<u>Aflatoxins</u> – fungi contaminants of peanuts, peanut products, tree nuts, pistachio nuts, pecans, walnuts, Brazil nuts, corn, corn products, milk, cottonseed; other grains and nuts less susceptible; carcinogen, IARC Group 1.
* C A	<u>Agar-agar</u> – 17, 32, 46, 51; natural vegetable gelatin that also functions as an

intestinal regulator; derived from marine algae; may cause flatulence, bloating, laxative effect; "animal studies show biochemical changes at very low doses where human health implications are not well understood;" used medicinally, may cause medical emergency if not taken with adequate amount of water.

C   Agave nectar – 49; most agave on the market is NOT traditional agave nectar which is high in nutrients with healing properties; agave nectar you buy in stores is actually syrup, not nectar; it's a highly processed sweetener; is about 90% fructose; may be from species linked to toxicity in humans; HFCS has also been labeled as agave syrup; may be processed with genetically engineered enzymes and caustic chemicals; may be contaminated with Hydroxymethylfurfural (HMF) which is potentially toxic, mutagenic and carcinogenic; not recommended.

X A   Aginomoto – see MSG.

C   Aguamiel – 49; sap of the agave plant; 90% fructose; see agave nectar.

X A   Ajinomoto – see MSG.

S A   Albumins – 46, 50, 51; may be egg, milk based.

* C   Alfalfa – 27; GMO alfalfa is currently being planted; choose only organic.

* C   Alginates – 17, 28, 46, 50; possible pregnancy complications; see alginic acid.

* C   Alginic acid – 5, 46; derived from brown seaweed; may cause eye, skin, respiratory, gastrointestinal irritation, birth defects;

toxicological properties not fully investigated.

X Alitame – 11; related to aspartame; not adequately tested; approved for use in Australia, New Zealand, Mexico, China; petition for approval in U.S. withdrawn in 2008.

X A Alkyl gallate – may cause liver problems.

X Alkyl sulfates – may cause skin rashes; may contain numerous chemicals; appears manufacturer trying to hide actual ingredients.

* S Allspice – 27.

X Allyl acetate – 9, 27; synthetic; toxic if swallowed; may be fatal if inhaled; harmful if absorbed through skin; causes lung damage; eye, skin, respiratory irritant; not adequately tested.

C Allyl anthranilate – 9, 27; synthetic; may be harmful if inhaled, absorbed through skin, swallowed; respiratory, skin, eye irritant; considered nontoxic, but not adequately tested.

C Allyl butyrate – 9, 27; synthetic; toxic if ingested, absorbed through skin; harmful if inhaled; eye, skin, respiratory irritant; not adequately tested.

C Allyl cinnamate – 9, 27; synthetic; harmful if swallowed, inhaled, absorbed through skin; eye, skin, respiratory irritant; not adequately tested.

C Allyl crotonate – 9,27; synthetic; harmful if swallowed, inhaled, absorbed through skin; eye, skin, respiratory irritant; not adequately tested.

X		Allyl cyclohexane acetate – 9, 27; see Allyl acetate.
C		Allyl cyclohexane butyrate – 9, 27; see Alllyl butyrate.
X		Allyl cyclohexane hexanoate – 9, 27; safety not determned.
X		Allyl cyclohexyl propionate – 9, 27; harmful if swallowed; not adequately tested.
X		Allyl cyclohexyl valerate – 9, 27; safety not determined.
X		Allyl hexanoate – 9, 27; safety not determned.
* C		Allyl isothiocyanate – 9, 27; toxic; may cause skin problems; causes cancer in rats; insufficient data to evaluate as a human carcinogen, IARC Group 3.
C A		Allyl sulfide – may cause breathing difficulty, kidney, liver problems.
* S A		Almond oil – 27; distilled to remove toxic hydrocyanic acid; nontoxic if distilled and free from hydrocyanic acid (prussic acid).
C		Aloe extract – healing and therapeutic benefits not well studied; aloe leaf juice is a laxative, other internal use benefits not substantiated; aloe gel has external benefits for wound healing, not properly studied for internal benefits; may cause gastrointestinal distress, kidney problems.
X		Alpha-acetolactate decarboxylase – enzyme derived from genetically modified bacteria.
X		Alpha-amulase – enzyme derived from genetically modified bacteria.
* C A		Alpha tocopherol – vitamin E; may be corn, peanut, soy based; see alpha tocopherol acetate.

C	<u>Alpha tocopherol acetate</u> – vitamin E; large doses may be harmful if high blood pressure; see nutrient additives.
X	<u>Aluminum</u> – may be associated with senility, memory problems, kidney problems, neurological problems, mouth ulcers, mineral malabsorption; not adequately tested.
* X	<u>Aluminum ammonium sulfate</u> – 4; may cause vomiting; see aluminum, ammonium.
* X	<u>Aluminum hydroxide</u> – 33; may cause constipation; see aluminum.
* X	<u>Aluminum potassium sulfate</u> – may cause gastrointestinal distress; see aluminum.
* X	<u>Aluminum sodium sulfate</u> – 14, 23, 39; harmful if swallowed, inhaled or on skin contact; see aluminum; not adequately tested.
* X	<u>Aluminum sulfate</u> – skin irritant; moderately toxic if swallowed; not shown to be safe; see aluminum.
C	<u>Amasake</u> – 49; fermented; may contain aspergillus oryzae; see fermented, aspergillus oryzae.
C	<u>Amino acids</u> – building blocks of protein; may be a hidden source of MSG.
X	<u>AminoSweet</u> – see aspartame.
X A	<u>Ammonia</u> – corrosive; toxic if inhaled; eye and mucous membrane irritant; can burn eyes, skin; can cause permanent damage; may cause mouth ulcers, nausea, kidney, liver problems; classified as hazardous by OSHA; on the EPA Extremely Hazardous

Substances list; best to avoid all products containing ammonia or ammonium salts.

* X   <u>Ammonium alginate</u> – see ammonia, alginates.

* X   <u>Ammonium bicarbonate</u> – 33; may cause gastrointestinal distress; see ammonia.

* X   <u>Ammonium carbonate</u> – see ammonia.

X   <u>Ammonium carrageenan</u> – see ammonia, carrageenan.

X A   <u>Ammonium caseinate</u> – 50; see ammonia, casein.

* X   <u>Ammonium chloride</u> – may cause gastrointestinal distress; toxic if ingested in large amounts; may cause irreversible damage; see ammonia.

* X   <u>Ammonium citrate</u> – 23, 26, 44; see ammonia.

* X A   <u>Ammonium gluconate</u> – 19; see ammonia.

X   <u>Ammonium hydrogen carbonate</u> – ammonium bicarbonate; see ammonia.

* X   <u>Ammonium hydroxide</u> – see ammonia.

* X   <u>Ammonium isovalerate</u> – see ammonia.

* X   <u>Ammonium phosphate</u> – see ammonia phosphates.

X   <u>Ammonium saccharin</u> – see saccharin.

* X   <u>Ammonium sulfate</u> – 18; see ammonia.

C   <u>Amyl acetate</u> – 27; may cause central nervous system depression, headaches, fatigue, mucous membrane irritation.

X   <u>Amyl alcohol</u> – 27; highly toxic, has caused deaths.

C A   <u>Amylases</u> – 18; may be soy based, genetically modified.

X   <u>Androstenedione</u> – steroid hormone; produced by the body, precursor to male and

	female hormones; used as supplement; androgenic and estrogenic; probably carcinogenic, IARC Group 2A; banned 2005; aka andro.
* C A	Anethole – 9; may cause mouth ulcers, burning sensation in mouth.
* C A	Angelica – may cause light sensitivity.
C	Animal or vegetable shortening – associated with heart disease, hardening of arteries, elevated cholesterol levels.
* S	Anise – herb, essential oil; healing properties; see essential oils.
φ C A	Annatto – 27, 38; may cause adverse skin, gastrointestinal, lung and central nervous system reations; headaches, restlessness, irritability in children; delayed sensitivity reactions.
C	Antioxidants (vitamin E) – may be genetically modified.
* S	Apricot kernel oil
C A	Arabinogalactan – 12, 19, 46, 51; not adequately tested.
X	Areca nut – carcinogen, IARC Group 1.
X	Aristolchia – herbal remedies containing plant species of the genus Aristolochia are carcinogenic, IARC Group 1.
X	Aristolochic acid – probable carcinogen, IARC Group 2B; toxic to kidneys; sometimes used in weightloss herbal remedies, also supplements for intestinal, immune problems and cough; aka Aristolchia, Bragantia, Asarum.
S	Arrowroot – no known toxicity.
X	Artemisia – 27; may cause headaches, nervous system irritation, gastro-intestinal

	distress, coma, death in large amounts.
* S	Asafoetida – herb, spice; flavor enhancer; digestive aid; healing properties; eaten raw causes digestive disturbances.
X A	Artificial color FD & C, U.S certified food color – contribute to hyperactivity in children; may contribute to learning and visual disorders, nerve damage; may be carcinogenic; see FD&C Colors.
X A	Artificial flavoring – may contain MSG or HVP; may be genetically modified; may contain chemicals designed to change your brain chemistry; may cause reproductive disorders, developmental problems; not adequately tested.
X	Artificial sweeteners – associated with health problems; see specific sweetener.
X	Asarum – see Aristolochic acid.
C A	Ascorbates – 7; may be corn based; may be genetically modified; see ascorbic acid.
* C A	Ascorbic acid – 1, 7; synthetic vitamin C; see nutrient additives; can enhance mineral absorption, can inhibit nitros-amine formation; may be corn based; may be genetically modified if manufac-tured in the U.S; combines with sodium benzoate to form carcinogen, benzene.
* C	Ascorbyl palmitate – 7, see ascorbic acid.
X	Aspartame – 11; may cause brain damage in phenylketonurics; may cause central nervous system disturbances, menstrual difficulties; may affect brain development in unborn fetus; 2005 study links low doses of aspartame to leukemia and lymphoma in rats; more research needed relating to

        aspartame's potential for causing cancer and brain tumors; genetically modified; aka Nutrasweet, Equal, Equal-Measure, Benevia, Spoonful, Canderel, AminoSweet.

C A   <u>Aspergillus oryzae</u> – 29; used in fermenting soy foods; may produce mycotoxins.

X A   <u>Autolyzed yeast</u> – contains MSG.

X A   <u>Azo dyes</u> – 10; may cause gastro-intestinal distress, nausea, hay fever, itching, high blood pressure; see artificial color....

S   <u>Azulene</u> – naturally occuring plant compound found in chamomile tea.

S   <u>Bakers yeast glycan</u> – 19, 46, 51.

C   <u>Bakers yeast protein</u> – 33, may contain MSG.

C A   <u>Baking powder</u> – 18, may contain corn; double acting may contain aluminum; may be genetically modified.

* S   <u>Baking soda</u> – see sodium bicarbonate.

C   <u>Banana oil</u> – see amyl acetate.

C   <u>B-Apo-8'-carotenal</u> – 38; used in candy, poultry feed to make egg yolks appear darker; restricted usage; not adequately tested.

C   <u>Barbados molasses</u> – 49; see sucrose.

C   <u>Barley malt</u> – 49; 1 Tbsp. contains 8 grams of sugar; better tolerated by people with blood sugar disorders; may be HFCS mislabeled as barley malt; processed sweetener; all sweeteners best avoided; may contain free glutamates, MSG; see sucrose, MSG.

* S   <u>Basil</u> – herb.

* S   <u>Bay leaves</u> – herb.

C   <u>Bee pollen</u> – may come from genetically

modified sources of pollen if hives near GMO crops.

* S A Beeswax – 19, 27, 29; white beeswax is bleached; yellow beeswax not bleached.

φ S Beet powder – 38.

X Benefat™ – see salatrim.

X Benevia – see aspartame.

* C Bentonite –51; see aluminum silicate.

* X A Benzaldehyde – 9; harmful if swallowed, inhaled, absorbed through skin; eye, skin, mucous membrane irritant; may cause central nervous system depression, decreased sex drive, immune system stress; possible mutagen; highly toxic; causes cancer in mice.

* X A Benzoate of soda – 40; can cause skin rashes, gastrointestinal upset, hyper-activity in children, neurological disorders; has caused birth defects in lab animals; a poison; moderately toxic if swallowed; may form benzene, a carcinogen, in presence of ascorbic acid (vitamin C); those with asthma or liver problems should avoid; aka sodium benzoate.

X Benzene – can form in products with sodium benzoate and ascorbic acid, i.e. vitamin C; IARC Group 1 carcinogen.

* C A Benzoic acid – 6, 40, see benzoate of soda.

* C Benzoic aldehyde – 9; see benzalde-hyde.

* X A Benzoyl peroxide – 13, 18; corrosive; destroys vitamin A, C, E, may cause skin rashes; inadequate evidence to classify as a carcinogen, IARC Group 3; banned in Europe.

C Benzyl acetate – 27; may cause

gastrointestinal upset; causes cancer in mice; no data available to evaluate if carcinogenic to humans, IARC Group 3

C   <u>Benzyl acetoacetate</u> – 27; see benzyl acetate.

X   <u>Benzyl alcohol</u> – 9; severe eye irritant; moderate skin and mucous memebrane irritant; may cause diarrhea, vomiting; poison if ingested.

C   <u>Benzyl ethyl ether</u> – 27; narcotic in large amounts.

C   <u>Benzyl formate</u> – 27; narcotic in large amounts.

* C1   <u>Bergamot</u> – 27; essential oil; anti-bacterial; strong photosensitizer; may cause contact dermatitis; see essential oils.

* C   <u>Beta Carotene</u> – 38; precursor to vitamin A; see nutrient additives.

S   <u>Beta glucans</u> – acomponent of cellulose.

X   <u>Beta-glucanase</u> – enzyme derived from genetically modified bacteria.

S   <u>Betacyanin</u> – natural color derived from beet.

X   <u>BETANAT®</u> – genetically engineered beta carotene.

S   <u>Betaxanthin</u> – natural color derived from beet.

X   <u>Betel nut</u> – carcinogen, IARC Group 1.

* X A   <u>BHA</u> – 7, 40; can cause liver and kidney damage, behavioral problems, infertil-ity, weakened immune system, birth defects, cancer; should be avoided by infants, young children, pregnant women and those sensitive to aspirin; possible carcinogen, IARC Group 2B.

* X A   <u>BHT</u> – 7; see BHA; limited data showing

	carcinogenicity in animals, no data available to evaluate if carcinogenic to humans, IARC Group 3; mutagen; may cause adverse reproductive effects.
X	Bitter almond oil – 9, 27; essential oil; human poison if ingested; skin, eye irritant; may cause diarrhea, nausea, vomiting, lung congestion, convulsions; may be contaminated with hydrogen cyanide.
C	Blackstrap molasses – 49; 1 Tbsp. contains 11-15 grams of sugars; contains small amounts of minerals, but is still 65% sucrose.
* S	Bicarbonate of soda – see sodium bicarbonate.
* S	Bile salts – used in digestive enzyme supplements and dried egg whites.
* C	Biotin – B vitamin; see nutrient additives.
C	Biphenyl – may cause nausea, vomiting, eye, nose irritation.
S	Blackberry bark extract – natural flavor.
X A	Blue No. 1 – see FD&C Blue No. 1.
X A	Blue No. 2 – see FD&C Blue No. 2.
X	Borax – illegal for use in foods, fruits and vegetable waxes; may be used in meat exports.
X	Boric acid – highly toxic; ingestion and topical application have caused poisoning.
C	Borneol – 27; may cause digestive disturbances, dizziness, convulsions.
X	Bouillon – may contain MSG.
X	Bragantia – see Aristolochic acid.
S	BriesSweet™ Brown Rice Syrups – enzyme produced; may be organic or non-organic; company mission is to provide

40

	healthier ingredients.
S	BriesSweet™ Tapioca Syrups – enzyme produced from tapioca starch; may be organic or non-organic.
S	BriesSweet™ Tapioca Maltodextrin - enzyme produced from tapioca starch; may be organic or non-organic.
CA	Brewer's yeast – by-product of brewing beer; contains some B vitamins, protein, various minerals; may cause abdominal discomfort, flatulence, headaches; long-term safety not known; caution during pregnancy and breast feeding; may cause itching and swelling in those allergic to yeast; may aggravate Crohn's disease; may interfere with medications for diabetes, depression and fungal infections; should not be taken by those with blood sugar problems, candida, high uric acid levels.
S	BriesSweet™ - 12, 21, 22, 46; derived from tapioca starch; organic or conventional; non-GMO.
X	Bromated flour – see potassium bromate.
X A	Brominated vegetable oil - 19, 24; contains bromine, a corrosive and toxic chemical, on the EPA Extremely Hazardous Substances list; has caused death in lab animals; stored in body fat; linked to major organ system damage, birth defects, growth problems; FDA restricts use as interim food additive; banned for soft drinks in India.
S	Bromelain – protein-digesting enzyme derived from pineapple.
C	Broth – may contain MSG.

- * C   <u>Brown algae</u> – Possible mercury contamination; avoid during pregnancy; not adequately tested.
- C   <u>Brown rice syrup</u> – 49; 1 Tbsp. contains 10 grams of sugars; better tolerated by people with blood sugar disorders; processed sweetener; all sweeteners are best avoided; may be a source of hidden MSG; see sucrose.
- * X   <u>Butane</u> – 41; may cause drowsiness, asphyxiation; mildly toxic if inhaled; neurotoxic in high doses; NIOSH says animal carcinogen.
- * C   <u>Butanoic acid</u> – 27; component of rancid butter; eye, skin, respiratory, gastrointestinal irritant; corrosive; harmful if swallowed.
- X   <u>Butyl acetate</u> – 9; may cause eye, skin, respiratory tract irritation; central nervous system depressant; toxic.
- X   <u>Butyl alcohol</u> – harmful if swallowed, inhaled, absorbed through skin; skin, eye, respiratopry tract irritant; harmful to central nervous system; potentially harmful to kidneys, liver.
- X   <u>1,3 butylene glycol</u> – 24, 32; skin and eye irritant; may cause nausea, vomiting, loss of consciousness if ingested; chronic overexposure may cause kidney or liver damage, nervous system disorders, death.
- * X A   <u>Butylated hydroxyanisole</u> – see BHA.
- * X A   <u>Butylated hydroxytoluene</u> – see BHT.
- X A   <u>Butylparaben</u> – 6, 40, skin irritant; associated with numerous health problems, including contact dermatitis and asthma;

	strong allergen; endocrine disruptor.
X	Butyraldehyde – 9; corrosive; harmful if swallowed, inhaled, absorbed through skin; extremely destructive of skin, eyes and mucous membranes; high exposure may cause accumulation of fluid in lungs; inhalation may be fatal.
C	Butyric acid – 27; see butanoic acid.
X A	BVO – see brominated vegetable oil.
* C	Cacao – source for cocoa, chocolate, cocoa butter; contains theobromine, a stimulant similar to caffeine, but less potent.
* X	Caffeine – 47; psychoactive, addictive drug; may cause headaches, irritability, fertility problems, increases risk of miscarriage, birth defects, low birth weight, heart disease, depression, nervousness, behavioral changes, insomnia, etc., inhibits fetal growth: insufficient data to evaluate as a human carcinogen, IARC Group 3.
* C	Calcium acetate – 44; low oral toxicity.
* C	Calcium alginate – 50; see alginates.
* C A	Calcium ascorbate – 7; see ascorbic acid.
C A	Calcium benzoate – see benzoate of soda.
C	Calcium bromate – 34; see potassium bromate.
* C	Calcium carbonate – 2, 24, 39, 52; see nutrient additives; may constipate.
X A	Calcium caseinate – 50; contains free glutamates, MSG; see casein, MSG.
* C	Calcium chloride – 23; see nutrient additives; may cause heart problems, gastrointestinal upset.
* C A	Calcium citrate – 14, 18, 44; may interfere with results of medical lab tests, see nutrient

	additives; may be corn based; may be genetically modified.
X	Calcium cyclamate – 11; banned in 1969 in U.S.; allowed "in products complying with drug provisions of the law."
* C	Calcium diacetate – 44; low oral toxicity.
C A	Calcium disodium EDTA – 7, 44; see EDTA.
C	Calcium formate – may cause urinary tract problems.
C A	Calcium fumarate – 1; may contain corn; may be genetically modified; see nutrient additives.
* C A	Calcium gluconate – 14, 23, 26, 44; may cause gastrointestinal upset, heart problems; may be corn sugar based; may be genetically modified.
* C	Calcium glycerophosphate – see nutrient additives.
* C	Calcium hexametaphosphate – 19, 44, 50; see calcium phosphate.
* C	Calcium hydroxide – 23; lye; skin irritant.
* C1A	Calcium iodate – 17, 18; caution if thyroid problems.
* C A	Calcium lactate – 3, 14, 23, 52; may cause cardiac or gastrointestinal disturbances; may be corn or milk based; see nutrient additives.
* C	Calcium orthophosphate – see calcium phosphate.
* C	Calcium oxide – 1, 18, 52; skin, mucous membrane irritant; see nutrient additives.
* C	Calcium pantothenate – vitamin B5; see nutrient additives.
C	Calcium peroxide – 13, 18; skin, eye,

	respiratory irritant; may cause gastro-intestinal distress; nontoxic if ingested.
* C	Calcium phosphate – 13, 18, 50; see nutrient additives; may reduce mineral absorption; may cause kidney damage.
* C A	Calcium propionate – 6, 40; can trigger behavioral changes.
* C	Calcium pyrophosphate – see nutrient additives.
C	Calcium silicate – 4; may cause kidney problems, possibility of asbestos contamination.
X A	Calcium sodium EDTA – 7, 44; see EDTA.
* C	Calcium sorbate – 6, 40; see sorbic acid.
* C A	Calcium stearate – 4; may be corn, peanut, soy based; may be derived from hydrogenated oils.
C A	Calcium stearoyl lactylate – 18, 19, 46; may be corn, milk, peanut, soy based; may be derived from hydrogenated oils; may be genetically modified.
* C	Calcium sulfate – 18, 34; may constipate; can kill rodents; see nutrient additives.
* S	Calendula – 27; herb.
* Cl	Camomile – 27; herb; see chamomile.
X A	Camphor oil - 27; has caused fetal death in pregnant women; toxic in adults.
* S A	Cananga – 27; essential oil; too much can cause headache or nausea; see essential oils.
* S	Candelilla wax – 29.
X	Canderel – see aspartame.
* X	Canola oil – becomes rancid easily; baked goods made with canola oil become moldy quickly; originally from hybridized rape seed, now most is genetically engineered

(75%); processed at extremely high temperatures; contaminated with solvent used for extraction; refined, bleached, degummed and deodorized at high temperatures; omega 3's coverted to trans fats; much canola oil used in processed foods is hydro-genated; depletes body stores of vitamin E; causes heart problems, blood platelet changes, retarded growth, shortened life span in lab animals; no long-term studies done on humans for safety.

φ X   Canthaxanthin –38; ingestion can cause night blindness, aplastic anemia; has caused death.

* C   Capers – bud of caper shrub; typically pickled; see pickled vegetables.

X   Caprenin® – 21; derived from hydro-genated rapeseed oil; not adequately tested.

* S   Caprylic acid – 9.

* S   Capsicum – may cause gastrointestinal upset.

* X   Caramel – 27, 38; may be derived from corn syrup; may be genetically modified; may be processed with caustic chemicals, sulfites or ammonia; may cause laxative effect; may be contaminated with carcinogen 4-MI aka 4-methylimidazole, not adequately tested.

* X   Caramel color – see caramel.

X   Caramel color III – processed with ammonia; causes reduced white blood count in lab animals and decreased immune functon.

* S   Caraway – herb; helps colic, digestive disorders, appetite and eliminate worms.

C A   Carbohydrate gum – resins derived from the

	bark of plants; see acacia gum, guar gum, gum ghatti, gum karaya, gum tragacanth.
* C	Carbon dioxide – 41; eye, skin, respiratory irritant; may cause lung, heat, central nervous system damage.
* C A	Carboxymethylcellulose – 4, 17, 19, 32, 46, 50, 51; mild irritant to eyes, skin, digestive and respiratory tract; may be harmful if inhaled or swallowed; chronic exposure causes cancer and adverse reproductive effects in rats (Sigma-Aldrich MSDS); toxological properties not fully investigated.
* S	Cardamom – herb/spice; therapeutic benefits.
* S	Cardamon – see cardamom.
* S	Cardamom oleoresin – see cardamom.
* S	Cardamom seed – see cardamom.
φ C A	Carmine – 38; may cause hives, life-threatening allergic reactions; derived from insects; not adequately tested.
* C	Carnauba wax – 29; derived from carnauba palm; refined and bleached; not considered harmful; toxicology not fully investigated.
* S	Carob bean – herb; therapeutic benefit; healthy alternative to chocolate; no caffeine or theobromine.
* S	Carob bean extract – see carob bean.
* S	Carob bean gum – 18, 51; extract from carob bean.
* C	Carotene – 38; see beta carotene, nutrient additives.
X	Carrageenan – 19, 46, 51; undegraded or native carrageenan has not caused cancer in test animals; inadequate information to determine carcinogenicity in humans, IARC

	Group 3; degraded carrageenan has caused cancer in rats; possible human carcinogen, IARC Group 2B; product labels do not distinguish between degraded and undegraded carrageenan; may contain free glutamates, MSG; should not be given to infants; see MSG; on FDA list for further study.
S	<u>Carrot oil</u> – 38; see essential oils.
C	<u>Carvacrol</u> – 9; essential oil; oregano oil; antimicrobial; therapeutic in very small amounts; large amounts are corrosive and harmful if swallowed or inhaled and destructive of skin and mucous membranes; can cause circulatory, respiratory depression, heart failure; do not use if pregnant or nursing; see essential oils.
* C	<u>Carvol</u> – 9, 47; see d-carvone.
* C	<u>d-carvone</u> – 9; may be toxic in large amounts.
* C	<u>l-carvone</u> – 9; see d-carvone.
* C	<u>Cascarilla bark</u> – herb; stimulant; possible narcotic effects.
* C A	<u>Casein</u> – 36, 50, 51; harmful to anyone with milk allergies; may contain traces of LAL, a chemical of questionable safety.
C A	<u>Caseinates</u> – cause kidney damage in rats.
* C1A	<u>Cassia</u> – herb/spice beneficial for lowering blood sugar, blood pressure, cholesterol, triglycerides; should not be used by women who are pregnant, nursing or who experience excessive menstrual bleeding; contains coumarin, avoid large amounts.
* C1A	<u>Cassia bark</u> – see cassia.
* C1A	<u>Cassia oil</u> – 9; can cause upper respiratory

	irritation; see cassia.
C1A	Castor oil – 9; therapeutic benefits; large amounts may cause pelvic congestion and induce abortion; should not be used if pregnant or nursing.
X	Catalase – enzyme derived from genetically modified fungi.
* C	Cayenne pepper – see capsicum.
C1	Cedar - 27; see cedarwood oil.
C	Cedar leaf oil – may cause reproductive failure, light sensitivity; see essential oils.
C1	Cedarwood oil – 27; may cause sensitivity to light; use cautiously or avoid if pregnant; see essential oils.
* C1A	Celery seed – herb; therapeutic benefit; diuretic; may cause photosensitivity; should not be used if pregnant; may cause fatal anaphylactic shock in those with celery allergy.
C	Cellulose – plant fiber; may be derived from cotton; may be GMO.
* C A	Cellulose gum – see carboxymethyl-cellulose.
C	Cellulose powder – see cellulose.
C	Cetyl alcohol – derived from petroleum or plant oils; skin, eye irritant; slightly hazardous if ingested or inhaled.
* C1A	Chamomile – 27; herb; anti-inflammatory; may interact with certain medications.
* S	Chervil – herb.
* S	Chervil extract – herb.
C	Chewing gum base – can be made from chicle, but most is petroleum based.
* S	Chicory – herb; therapeutic benefit.

X		Chlorine dioxide – 13, 34; poison; severe respirtory and eye irritant; may cause headache, shortnessof breath, wheezing, cough, vomiting; emphysema; bronchitis, pulmonary edema; intense irritation at 5 ppm; classified as a pesticide by the EPA; chlorine is on the EPA Extremely Hazardous Substances list; unstable; highly flammable; may explode on impact; not tested for carcinogenic effects.
C		Cholecalciferol – vitamin D3; see nutrient additives.
* C		Choline bitartrate – a vitamin; see nutrient additives.
* C		Choline chloride – a vitamin; see nutrient additives.
* X		Chondrus extract – 46; see carrageenan.
* X		Chymosin – component of rennet used in cheese making; most is genetically engineered.
X		Chymostar – genetically modified rennet for making cheese.
C A		Cinchona bark – 27; contains quinine; stimulates digestion; medically used to treat malaria; can cause death in large amounts.
X A		Cinnamal – see cinnamic aldehyde.
* X A		Cinnamaldehyde – 9; see cinnamic aldehyde.
* X A		Cinnamic aldehyde – 9; skin sensitizer; toxin; neurotoxin; immunotoxin; gastrointestinal, mucous membrane and skin irritant; may cause depigmentation.
* C A		Cinnamon – see cinnamon bark extract.
* C1		Cinnamon bark extract – essential oil; antibacterial; anti-viral; anti-fungal;skin

		sensitizer; may cause skin rashes; avoid if pregnant; see essential oils.
* C1	Cinnamon bark oil	– essential oil; antibactrial; anti-viral; anti-fungal; skin sensitizer; may cause skin rashes, light sensitivity; causes contact dermatitis; avoid if pregnant; see essential oils.
X A	Cinnamyl aldehyde	– see cinnamic aldehyde.
* C	Citral	– 9; interferes with wound healing; causes contact dermatitis; sensitizer.
* C A	Citric acid	– 1, 7, 27, 44; generally produced by fermentation by yeast or mold; solvent extracted; may erode tooth enamel; may be corn based; may be genetically modified; may contain free glutamates, MSG.
C	Citric acid ester of mono- and di-glycerides	– soy lecithin substitute; manufacturer claims non-GMO and free of hydrogenated fats; not adequately tested.
* C A	Citronella	– essential oil; antibacterial; anti-fungal; anti-inflammatory; antiseptic; may irritate sensitive skin; skin sensitizer if used frequently; inhalation of pure oil can increase heart rate; use cautiously or avoid if pregnant; see essential oils.
C1	Citrus bioflavonoids	– vitamin P complex; pregnant women should not take megadoses; link between infant leukemia and high doses in mother; see nutrient additives.
* S	Citrus peel	– 27.
X A	Citrus Red No. 2	– 10; monoazo dye; used to color orange skins; possible carcinogen,

51

		IARC Group 2B; restricted to specific uses; see artificial color....
* C1	Clary – see clary sage.
* C1	Clary sage – herb, essential oil; therapeutic benefit; sedative, do not use before driving; avoid if pregnant.
* C1	Clove bud oil – 27; healing properties; may cause contact dermatitis, sensitization; has caused gastrointestinal irritation in lab animals; use cautiously or avoid if pregnant; see essential oils.
* C1	Clove leaf oil – 27; see clove bud oil.
* C1	Clove stem oil – 27; see clove bud oil.
* C	Clover – 27; may cause sensitivity to light.
* C1A	Cloves – 9; see cassia oil.
* C A	CMC – see carboxymethylcellulose.
X A	Coal tar dyes – 10; mostly derived from petroleum; may cause hay fever, skin rashes, nausea, itching, gastrointestinal distress, high blood pressure; coal tar pitches and coal tars are classified as Group 1 carcinogens by IARC; see artificial color....
C	Cobalamin (vitamin B12) – see nutrient additives; may be genetically modified.
X	Cobalt and cobalt compounds – part of vitamin B12, beneficial for health in small amounts; high levels may cause dermatitis, adverse effects on heart, lungs, kidneys, liver; has caused cancer in lab animals; possible carcinogen, IARC Group 2B.
* C	Coca (decocanized) – 27; raw material from which cocaine is produced, although the cocaine alkaloid has been removed from the coca approved as a food additive.

φ C A	Cochineal	– 38; see carmine.
* C	Cocoa	– contains caffeine-like chemical; Dutch process cocoa processed with alkali; see caffeine.
* S	Cococin™	- freeze dried coconut water solids; supports cell growth.
* S	Coconut oil	– helps the body metabolize fatty acids; substitute for butter; use for frying and baking; use only unrefined, non-hydrogenated, non-bleached, non-deodorized.
* X	Coffee	– contains naturally occuring carcinogens; combines with stomach acid to form a potent toxin; see caffeine.
* X	Cola nut	– 27; essential oil; contains caffeine, carcinogenic N-nitroso and tannin compounds.
X	Cold pasteurization	– food irradiation.
C	Condensed milk	– may be genetically modified.
* C A	Colorose	– 49; see invert sugar.
* C	Confectioner's glaze	– 30; no studies evaluating safety in food use.
C	Confectioner's sugar	– see sucrose; may be genetically modified.
* C	Copper carbonate	– see copper salts.
* C	Copper chloride	– see copper salts.
* C	Copper gluconate	– see copper salts.
* C	Copper hydroxide	– see copper salts.
* C	Copper orthophosphate	– see copper salts.
* C	Copper oxide	– see copper salts.
* C	Copper pyrophosphate	– see copper salts.
C	Copper salts	– see nutrient additives; skin and mucous membrane irritants; can cause vomiting.

* C		Copper sulfate – most highly irritating copper salt; see copper salts.
* X		Corn dextrin – hydrolyzed corn starch; may contain MSG; may be GMO (60%).
* X A		Corn gluten – 42, 50; may be genetically modified (60%).
C		Corn masa – may be GMO.
C		Cornmeal – may be GMO (60%).
X		Corn oil – commonly GMO (60%); causes cancer in rats.
X		Corn protein – hydrolyzed protein; contains MSG; see free glutamates, MSG, hydrolyzed vegetable protein; may be GMO (60%).
* C1		Corn silk – 27; herb; consult health care professional before using if pregnant.
* X A		Corn starch – 51; may cause hay fever, eye, nose irritation; may be GMO (60%).
* X A		Corn sugar – see high fructose corn syrup; may be GMO (60%).
* X A		Corn syrup – 49, 51; associated with blood sugar problems, depression, fatigue, B-vitamin deficiency, hyperactivity, tooth decay, periodontal disease, indigestion; may be genetically modified (60%).
X A		Cottonseed oil – unrefined, used as a pesticide; highly refined for food use; frequently hydrogenated or partially hydrogenated; more than 50% omega-6 fatty acids; trace of omega-3's; may be highly contaminated with pesticide residue; frequently GMO (83%).
* C1		Cream of tartar – 1, 4, 33; may have laxative effect; caution if kidney or heart problems.

* C A   Croscarmellose sodium – see sodium carboxymethylcellulose.

C   Crospovidone – ingredient found in drugs and nutritional supplements; not considered a health hazard by the Joint FAO/WHO Expert Committee for Food Additives; polymer of N-vinyl-2-pyrrolidone, which is classified as an animal carcinogen by the American Conference of Governmental Industrial Hygenists (ACGIH); not adequately tested.

X   Crystalline fructose – pure fructose; derived from corn; may be contaminated with arsenic, lead, heavy metals; may cause cirrhosis; used in health drinks; may be GMO.

* C   Cumin – 27; skin irritant; moderately toxic on skin contact and if ingested; possible mutagen.

* C1A   Cuprous iodide – 17, 18; caution if thyroid problems.

C   Cyanocobalamin – vitamin B12; your body needs this nutrient in the proper amounts; taken in excess, can cause an excess of cobalt in your tissues and contribute to heart problems; excess can also depress the thyroid; see nutrient additives; may be genetically modified.

X   Cyclamates – 11; banned in U.S. in food; causes cancer in mice; insufficient evidence to determine if carcinogenic to humans, IARC Group 3

X   Cyclodextrin-glucosyl transferase – enzyme derived from genetically modified bacteria.

C   Cyclodextrins – may be harmful if

swallowed, absorbed through skin; eye, skin, lung irritant; prolonged use may cause liver, urinary tract damage; adverse reproductive effects in lab animals; may be derived using GMO enzymes; not adequately tested.

C   <u>Cysteine</u> – amino acid; may be genetically modified.

* C   <u>D-Pantothenyl alcohol</u> – see nutrient additives.

C   <u>Dairy-Lo®</u> – 21; fat substitute; see microparticulated protein product.

S   <u>Date sugar</u> – 49; 1 Tbsp. contains equivalent of 3 grams of sugars; all sweeteners best avoided; use sparingly.

* C A   <u>Datem</u> – see diacetyl tartaric acid esters of mono- & diglycerides.

C   <u>7-Dehydrocholesterol</u> – vitamin D3; see nutrient additives.

* C   <u>Decanal</u> – 9; moderately toxic if swallowed; eye and skin irritant.

C   <u>Devan Sweet</u> – 49; see rice syrup powder.

* C   <u>Dextrans</u> – very large doses may cause gastrointestinal discomfort; a carbohydrate rapidly converted to glucose; diabetics should avoid; has caused cancer in lab animals.

* C A   <u>Dextrin</u> – 17, 51; partially hydrolyzed starch; may be from wheat or corn; corn may be genetically modified.

* C A   <u>Dextrose</u> – 49; may be derived from corn; may be GMO; see glucose.

C A   <u>Dextrose anhydrous</u> – 49; derived from corn; may be genetically modified.

C A   <u>Dextrose monohydrate</u> – 49; derived from

      corn; may be genetically modified.
* X   <u>Diacetyl</u> – 9; has caused cancer in lab animals; may be genetically modified; not adequately tested.
* C A <u>Diacetyl tartaric acid esters of mono- & diglycerides</u> – 19; aka datem; see mono- & diglycerides.
* C A <u>Diacylglycerol</u> – see mono- & diglycerides; may be genetically modified.
* C   <u>Dibasic ammonium phosphate</u> – see ammonium, phosphates.
* C   <u>Dibasic calcium phosphate</u> – see calcium phosphate.
* C   <u>Dibasic potassium phosphate</u> – see potassium phosphate.
* C   <u>Dicalcium phosphate</u> – see calcium phosphate.
 X   <u>Dichlorvos</u> – pesticide used on produce, flea collars, food packaging; causes cancer in rats and mice; mutagen; EPA says possible carcinogen; on the EPA Extremely Hazardous Substances list.
 X   <u>Diethylene glycol</u> – poison; detected in wines (1985) and medications (as recently as 2006 in Panama); causes kidney failure and death when ingested; causes cancer in rats (Carcinogenic Potency Project).
* C A <u>Diglycerides</u> – see mono- & diglycerides; may be genetically modified.
 C   <u>Diindolylmethane</u> – stable breakdown product of indole-3-carbinole; evidence it helps prevent hormone-related cancers; consult health care practitioner regarding use.
* C   <u>Dilauryl thiodipropionate</u> – 7; short term

studies on rats, guinea pigs and dogs have shown no toxic effects; long term studies on rats show no toxic effects; no long term studies on other animals or humans; not adequately tested.

* S   Dill – 27; can cause sensitivity to light.
C1   Dill oil – 27; potentially toxic in large amounts; use cautiously if epileptic.
S   Dillseed – 27.
C   DIM – see diindolylmethane.
X   Dimethyl dicarbonate - 40, 46; skin, eye, respiratory, gastrointestinal irritant; may cause stomach pain; severe over-exposure can cause death; chronic effects on humans unknown; not adequately tested.
X   Dimethyl sulfate – used in the manufacture of flavorings; probable carcinogen, IARC Group 2A; on EPA Extremely Hazardous Substances list.
C   Dimethylpolysiloxane – 5; possibility of asbestos contamination; may cause kidney problems.
C   DIMODAN® NH Distilled Monoglyceride – new trans-fat free, hydrogenation free monoglyceride; made from edible palm-based fat; non-GMO; patent-pending technology; new ingredient for bakery foods; processed food ingredient.
X   Dioctyl sodium sulfosuccinate (DSS) – 19, 46; eye, mucous membrane and skin irritant; potentially harmful if absorbed through skin, inhaled or swallowed; laxative effect; may cause gastrointestinal irritation, birth defects; not adequately tested.

- X   Diphenyl – may cause nausea, vomiting, eye, nose irritation, liver damage; exposure to large amounts can cause convulsions, central nervous system depression, paralysis.
- * C   Dipotassium phosphate – 14, 44; may reduce mineral absorption; may cause kidney damage.
- X A   Disodium EDTA – may cause formation of carcinogens in products containing nitrogen compounds; mucous membrane, eye, skin irritant;may cause asthma, kidney damage; penetration enhancer; see EDTA.
- X A   Disodium guanylate – 26; can aggravate gout; may be soy or yeast based; used in products containing MSG; not adequately tested.
- X A   Disodium inosinate – 26; can aggravate gout; may be soy or yeast based; used in products containing MSG; not thoroughly tested.
- * C   Disodium phosphate – 12, 14, 19, 44; mild skin, mucous membrane irritant; ingesting large amounts may cause diarrhea, nausea, vomiting; see nutrient additives.
- * C   Disodium pyrophosphate – see disodium phosphate.
- C   Disodium riboflavin phosphate – vitamin B2; see nutrient additives.
- X   DMDC – see dimethyl dicarbonate.
- C   Dough conditioners – reduce mineral availability.
- C   DSS – see dioctyl sodium sulfo-succinate.
- * C   D-tagatose – see tagatose.

- * S   Dulse – 27; from seaweed.
- S   Durabrite® Colors - natural food colors.
- X A   Durkex oil – 24, 29; refined, bleached, deodorized, partially hydrogenated oil; see hydrogenated vegetable oil.
- X A   EDTA – 7, 44; may cause skin irritation, gastrointestinal upset, liver, kidney damage, mineral imbalances; may cause errors in results of medical lab tests; see disodium EDTA.
- X   Electronically pasteurized – irradiated.
- C   Enliten® - Rebaudoside A, aka Reb A; derived from stevia leaf; may be extracted with heat, chemicals; see Reb A.
- C A   Enzyme of aspergillus oryzae - 31; may contain MSG; may be GMO.
- * C   Enzyme-modified fats – 27; may contain free glutamates, MSG; see MSG, enzymes.
- X   Enzyme-modified soy protein – may contain MSG; may be GMO; see MSG, free glutamates, soy.
- C   Enzymes – may contain MSG; may be genetically modified.
- C   Epsom salts – see magnesium sulfate.
- X   Equal – 11; see aspartame.
- X   Equal Spoonful – 11; see aspartame.
- X   Equal-Measure – see aspartame.
- X   Ergocalciferol – vitamin D2; see nutrient additives; on the EPA Extremely Hazardous Substances list.
- * C A   Erythorbic acid – 7, 16; enhances non-heme iron absorption; eye, skin respiratory irritant; may be corn based; may be genetically modified.

- C   Erythritol – 26, 32, 44, 46, 49, 50, 51; may be derived from corn; better tolerated than other sugar alcohols; may be GMO; see sugar alcohols.
- X A   Erythrosine – may cause overactive thyroid, sensitivity to light; see coal tar dyes.
- C   Essential oils – properly extracted for highest purity and used knowledgeably, essential oils have great therapeutic benefit; used as an ingredient in food products, most essential oils are safe; used full strength, cautions may be advised; pure essential oils do not cause allergic reactions; reactions to pure essential oils are detox reactions. If synthetics are added, the effect may be altered; if solvent or heat extracted, they are highly refined and lose beneficial properties; toxic solvent extraction contaminants may remain in the finished product; all absolutes are solvent extracted. Different brands of oils vary in quality, effectiveness and safety; 95% of oils sold are altered and are perfume or food grade, not pure; if uncertain if synthetics are added, more caution is advised. Consult your physician if pregnant or other health issues.
- C   Ester gum – 19, 27, 45; solvent extracted; inadequate data on safety available; see glycerol ester of wood rosin.
- * C   Ethanal – 9; see acetaldehyde.
- C   Ethoxylated mono- & diglycerides – 18; contain trans fats; eye, skin, digestive irritant; not adequately tested.
- C   2-Ethyl-5-methylpyrazine – 27; may be

        harmful if ingested, absorbed through skin or inhaled; eye, skin, respiratory irritant; not adequately tested.

* X   Ethyl acetate – 9; skin irritant; nervous system depressant; prolonged inhalation can cause kidney, liver damage; may contribute to secondary infection.

* X   Ethyl alcohol – fatal in large doses; causes cancer in rats.

* C   Ethyl butyrate – 9; skin irritant; mildly toxic if ingested.

* C A   Ethyl cellulose – 12, 22; see carboxy-methylcellulose.

* C   Ethyl formate – 6, 27; skin and mucous membrane irritant, narcotic in high concentrations.

C   Ethyl heptanoate – 9; skin, eye, respiratory irritant; no long-term studies; toxicological effects not fully investigated.

C A   Ethyl maltol – 26; harmful if ingested; skin, respiratory irritant; not adequately tested.

C   Ethyl methyl phenylglycidate – 9; adversely affects nervous system in lab animals.

C A   Ethyl propionate – 9; see propionic acid.

* X   Ethyl vanillin – 9; severe skin and eye irritant; hazardous if ingested; may cause organ system damage, cancer in lab animals; moderately toxic.

X A   Ethylene glycol – hazardous in small amounts.

X A   Ethylenediamine tetraacetic acid – see EDTA.

φ X A   FD&C Blue No. 1 – 10; may cause itching, low blood pressure; carcino-genic in rats, no data on humans, IARC Group 3; not adequately tested; see artificial color....

φ X A   FD&C Blue No. 1 Lake – coal tar dye; may contain aluminum; see FD&C Colors, aluminum.

φ X A   FD&C Blue No. 2 – 10; hazardous if ingested or inhaled; skin, respiratory tract, gastrointestinal irritant; may cause diarrhea, nausea, vomiting; may cause itching, low blood pressure, brain tumors in lab animals; not adequately tested for ability to cause cancer, mutagenic or genotoxic effects; see artificial color....

φ X A   FD&C Blue No. 2 Lake – coal tar dye; may contain aluminum; see FD&C Colors, aluminum.

φ X   FD&C Colors – colors considered safe by the FDA for use in food, drugs and cosmetics; most of the colors are derived from petroleum and known as coal tar colors; must be certified by the FDA not to contain more than 10ppm of lead and arsenic; certification does not address any harmful effects these colors may have on the body; most coal tar colors are potential carcinogens, may contain carcinogenic contaminants, and cause allergic reactions; the "Lakes" are mixed with aluminum hydroxide.

φ X A   FD&C Citrus Red No. 2 – 10; possible carcinogen, IARC Group 2B; restricted to specific uses; see artificial color....

φ X A   FD&C Green No. 3 – 10; see FD&C Colors.

φ X A   FD&C Orange No. 2 – possible carcinogen, IARC Group 2B.

φ X A   FD&C Red No. 3 – 10; approved for food

and ingested drugs; not approved for cosmetics or externally applied drugs; causes thyroid tumors in lab animals; found to cause cancer by Oak Ridge National Laboratory in 1996.

φ X A   FD&C Red No. 40 – 10; monoazo color; may be contaminated with carcinogens; see artificial color....

φ X A   FD&C Red No. 40 Aluminum Lake – may be contaminated with a carcinogen; see FD&C Colors, aluminum.

φ X A   FD&C Yellow No. 5 – 10; may cause hay fever, gastrointestinal upset, skin rashes; avoid if aspirin sensitive; see coal tar dyes, artificial color....

φ X A   FD&C Yellow No. 5 Lake – may contain aluminum; see FD&C Yellow No. 5, aluminum.

φ X A   FD&C Yellow No. 6 – 10; causes tumors in lab animals; contaminated with carcinogens; see coal tar dyes, artificial color....

φ X A   FD&C Yellow No. 6 Lake – may contain aluminum; see FD&C Yellow No. 6, aluminum.

* S A   Fennel – herb.

* S   Fenugreek – herb.

C   Fermented – traditionally fermented foods have long been considered healthy; Weston A Price Foundation recommends lacto-fermented foods; TruthInLabeling.org says fermentation may cause free glutamic acid (MSG) to form; MSG sensitive may react.

* C   Ferric ammonium citrate – see nutrient additives, ammonium; not adequately

		tested.
*	C	<u>Ferric chloride</u> – see nutrient additives; not adequately tested.
*	C	<u>Ferric citrate</u> – see ferric chloride.
	C	<u>Ferric orthophosphate</u> – see nutrient additives; see ferric chloride.
*	C	<u>Ferric phosphate</u> – see ferric chloride, nutrient additives.
*	C	<u>Ferric pyrophosphate</u> – see nutrient additives, ferric chloride.
*	C	<u>Ferric sodium pyrophosphate</u> – see ferric chloride, nutrient additives.
*	C	<u>Ferric sulfate</u> – see ferric chloride, nutrient additives.
*	C	<u>Ferrous fumarate</u> – see nutrient additives, ferrous gluconate.
*	C	<u>Ferrous gluconate</u> – 14, 23, 26, 44; see nutrient additives; may cause gastrointestinal disturbances; not adequately tested.
*	C	<u>Ferrous lactate</u> – see nutrient additives, ferrous gluconate.
*	C	<u>Ferrous sulfate</u> – see nutrient additives, ferrous gluconate.
	X	<u>Flavorings</u> – may contain MSG; may be genetically engineered; see natural flavors.
	X	<u>Flavors</u> – may contain MSG; may be genetically engineered see natural flavors.
	S	<u>Florida crystals evaporated cane juice</u> – natural or certified organic unrefined sugar; minimally processed; use small amounts; sugar in any form is harmful if more than small amount used; see sucrose.
	X	<u>Fluoride</u> – skin, eye, nose, throat irritant; poison; interferes with hor-mones; may

affect thyroid function; causes premature aging, weakening of immune system, mottling of teeth, anemia, joint stiffness, calcified ligaments, genetic damage; associated with bone cancer in boys; "emerging neurotoxin;"ADA advises NO fluoridated water for babies under one year of age; IARC Group 3.

C   Folacin – folic acid; B vitamin; see nutrient additives.

C   Folic Acid – B vitamin; see nutrient additives.

* C   Food shellac – see confectioner's glaze.

* C   Food starch – may be genetically modified; see starch.

X A   Formaldehyde – 40; used in animal feed; carcinogen, IARC Group 1; causes DNA damage; on the EPA Extremely Hazardous Substances list.

X   Formic acid – 27; may cause urinary tract problems.

S   FOS – 49; see fructooligosaccharides.

X A   Free glutamates – 26; may cause brain damage, especially in children; may be genetically modified; **always found in** Accent, Ajinomoto, autolyzed yeast, calcium caseinate, calcium glutamate, gelatin, glutamate, glutamic acid, hydrolyzed corn gluten, hydrolyzed protein, hydrolyzed soy protein, Marmite, monopotassium glutamate, monosodium glutamate, monoammonium glutamate, magnesium glutamate, natrium glutamate, pea protein, plant protein extract, sodium caseinate, textured protein, TVP, Vegemite,

Vetsin, yeast extract, yeast food and yeast nutrient; **may be in** barley malt, boullion, broth, carrageenan, citric acid, enzymes, anything enzyme modified, anything fermented, flavors & flavorings, malt extract, malt flavoring, maltodextrin, natural flavors and flavorings, natural chicken flavoring, natural beef flavoring, natural pork flavoring, pectin, pre-basted poultry, protease, protease enzymes, seasonings, soy protein, soy protein concentrate, soy protein isolate, soy sauce, soy sauce extract, stock, Taste No. 5, Umami, whey protein, whey protein concentrate, whey protein isolate, anything that is enzyme modified, fermented, protein fortified or ultrapasteurized and foods that advertise NO MSG, NO Added MSG or NO MSG Added; see MSG, processed free glutamates.

C   Fructooligosaccharides – 49; highly refined; can be fructose mislabeled as FOS; may cause intestinal gas, bloating.

* X A   Fructose – 49; 1 Tbsp. contains 12-15 grams of sugars; may be refined corn syrup or beet sugar; may cause elevated uric acid, blood pressure, obesity , diabetes, liver damage; may cause gastrointestinal distress, elevated triglycerides; may cause loss of essential minerals from the body; large amounts have caused tumors in mice; may be genetically modified.

C A   Fruit juice concentrate – 49; 1 Tbsp. contains 5.5-8.5 grams of sugars; highly proessed; contains high concentrations of

fructose; may contain fungicides and
pesticides; see fructose, sucrose.

C   Fruit pectin – 19, 46, 51; may contain free glutamates.

* C A   Fumaric acid – 1, 27; eye, skin, respiratory irritant; in large amounts, may cause gastrointestinal distress, kidney, liver damage; may be corn based, most synthetically produced; not adequately tested.

C   Fungal protease from aspergillus oryzae – see enzyme of aspergillus oryzae.

C A   Furcelleran – 19, 46; inadequately tested.

C   Galactitol – see sugar alcohols.

C   Galangal – plant/herb in ginger family used in Thai cooking; therapeutic at low doses; digestive irritant above recommended doses.

* S   Garlic – 27.

* X A   Gelatin – 17, 46, 51; hydrolyzed collagen protein; contains MSG; may contain sulfur dioxide; cannot be determined if from BSE animals (i.e. animals with mad cow disease); processing destroys most BSE prions; FDA warning in 2004 re testicular cancer resulting from gelatin in Jell-O due to processing method used for short time; see sulfur dioxide, MSG.

* X A   Gelatin hydrosylate – hydrolyzed collagen; see gelatin.

X A   Gelatine – see gelatin.

C   Gellan gum – 46, 51; produced through bacterial fermentation (may be a source of hidden MSG); contains nitrogen-containing compounds; eye, skin, mucus membrane, respiratory irritant; harmful if inhaled,

* C   swallowed or absorbed through skin; may cause diarrhea; poorly absorbed; non-toxic in rat studies: toxological properties not fully investigated.
* C   Geranial – 9; see citral.
* C A   Glucono delta-lactone – 1, 33; may be corn based; may be genetically modified.
* C A   Gluconolactone – 1, 33; may be corn based; may be genetically modified.
C A   Glucose – 49, 51; may be corn based; safer than fructose; not recommended for diabetics or the insulin resistant; may be genetically modified.
X   Glucose isomerase – enzyme derived from genetically modified bacteria.
X   Glucose oxidase – enzyme derived from genetically modified fungi.
C   Glucuronolactone – safe in small quantities; there is no research on the safety of the large quantities in energy drinks; should not be consumed with alcohol; not adequately tested.
X A   Glutamates – 26; glutamate is an important chemical that relays electrical signals between nerve cells and other cells: too much can excite brain cells to death; glutamates in food may be genetically modified; see MSG.
* X A   Glutamic acid – 26, 43; non-essential amino acid; source of hidden MSG when used singly and out of balance with other amino acids as a result of food manufacturing process; may be GMO; see MSG, free glutamates, processed free glutamic acid.
* X A   Glutamic acid hydrochloride – 26; see

MSG.
- C A   <u>Gluten</u> – may be genetically modified.
- C   <u>Glycerides</u> – 5, 8, 17, 18, 19, 46; may be derived from animal fat, soy or canola oil, or be synthetically produced; always partially hydrogenated; contain trans fats; may be GMO.
- X   <u>Glycerides and polyglycides of hydrogenated vegetable oils</u> – hydrogenated vegetable oil reacted with polyethylene glycol; see hydrogenated vegetable oil, polyethylene glycol; may be genetically engineered.
- * C A   <u>Glycerin</u> – 8, 25, 32; may be derived from tallow, corn, peanut, soy or propylene alcohol; can be beneficial in delaying dehydration while exercising, it should be avoided by those with high blood pressure, kidney disease or diabetes; aka glycerol; corn and soy may be GMO; see glycerol.
- * C A   <u>Glycerine</u> – see glycerin, glycerol.
- C   <u>Glycine</u> – amino acid; body produces what you need; ingestion of large doses may cause nausea; mildly toxic; causes cancer in rats; may be GMO.
- * C A   <u>Glycerol</u> – 22, 32, 49, 51, also known as glycerin, glycerine, glyceritol, glycyl alcohol, propane-1,2,3-triol, 1,2,3-propanetriol, 1,2,3-trihydroxypropane; may be synthesized from animal or vegetable fats (corn or soy) or petroleum; may be GMO; see sugar alcohols.
- X   <u>Glycerol ester of gum rosin</u> – 19, 46; European Food Safety Authority states there is insufficient data to determine safety, FDA

	allows in food in U.S.
X	<u>Glycerol ester of tall oil rosin</u> – 19, 46; petroleum derivative; may be contaminated with lead; European Food Safety Authority states there is insufficient data to determine safety, FDA allows in food in U.S.
C	<u>Glycerol ester of wood rosin</u> – 19, 46, 51; solvent estracted; heated to very high temperatures; may be hydrogenated; no long term studies done on toxicity.
C	<u>Glycerol monooleate</u> – may be genetically modified; see mono- & diglycerides
C	<u>Glyceryl abietate</u> – see ester gum.
* C	<u>Glyceryl monostearate</u> – 19, 45; may be lethal to lab animals in large doses.
* C	<u>Glyceryl triacetate</u> – 27, 45; may be lethal to lab animals in large doses.
C	<u>Glycine</u> – may be genetically modified.
* C A	<u>Glycyl alcohol</u> – see glycerol.
* C	<u>Glycyrrhizin</u> – 27; see licorice.
C	<u>Golden syrup</u> – 49; may contain corn; may be genetically modified; see treacle.
C	<u>Granular fruit source</u> – 49; 1 Tbsp contains 7.5 grams of sugars.
X	<u>Grapeseed oil</u> – high in omega-6 fatty acids, which are best avoided; processed with carginogenic solvents, some of which remain in the oil.
X A	<u>Green No. 3</u> – 10; see FD&C Green No. 3.
* S	<u>Ground limestone</u> – 27.
* C A	<u>Guar gum</u> – 46, 51; may cause nausea, gastrointestinal upset, bloating.
C	<u>Guarana</u> – central nervous system stimulant; contains caffeine; see caffeine.
* C A	<u>Gum arabic</u> – see acacia gum.

S		<u>Gum benzoin</u> – 27; no known toxicity.
C		<u>Gum furcelleran</u> – 19, 46, 51; on FDA's list to be studied for mutagenic, teratogenic, subacute and reproductive effects; not adequately tested.
* C A		<u>Gum ghatti</u> – 19, 46; not adequately tested.
* C A		<u>Gum guaiac</u> – 7; not adequately tested.
* C A		<u>Gum karaya</u> – 19, 50; not adequately tested.
* C A		<u>Gum tragacanth</u> – 46, 51; can cause severe allergic reactions; not adequately tested.
* S		<u>Helium</u> – 41.
X		<u>Hemicellulase</u> – enzyme genetically modified from bacteria or fungi.
C		<u>Hemicellulose</u> – may be GMO.
X		<u>Heptanal</u> – see acetaldehyde.
X A		<u>Heptylparaben</u> – 6, 40; endocrine disruptor; pregnant women, children, aspirin sensitive should always avoid; not adequately tested; see parabens.
* X A		<u>HFCS</u> – see high fructose corn syrup.
* X A		<u>High fructose corn syrup</u> – see fructose, corn syrup; may be contaminated with mercury; may contribute to elevated cholesterol and formation of blood clots; interferes with action of white blood cells; may contribute to diabetic complications; contributes to obesity; may be GMO.
X A		<u>High maltose corn syrup</u> – see high fructose corn syrup, maltodextrin.
C		<u>Honey</u> – 49; 1 Tbsp. contains 16-18 grams of sugars; may contain botulism spores; potentially harmful or fatal to children under 18 months; may be from genetically modified sources of pollen if hives near GMO crops; there is no USDA organic

standard for honey, most organic honey is imported; USDA organic seal on honey is meaningless; use raw, unfiltered honey; use sparingly.

X A   HPP – see hydrolyzed vegetable protein.

C A   HSH – see hydrogenated starch hydrolysate.

X A   HVP – see hydrolyzed vegetable protein.

* C   Hydrochloric acid – 14, 39; may cause irritation to mucus membranes, gastrointestinal distress.

* C   Hydrogen peroxide – 13, 40; 3% solution is mild disinfectant and antiseptic; do not swallow; 3% food grade can be therapeutic in small amounts under the supervision of a qualified health care professional; most hydrogen peroxide on the market contains chemical stabilizers and should not be used internally; safety of food grade hydrogen peroxide depends upon the concentration; higher concentrations of any grade of hydrogen peroxide may cause DNA damage and cell death; IARC Group 3; on the EPA Extremely Hazardous Substances list for concentrations greater than 52%.

C A   Hydrogenated glucose syrup – see hydrogenated starch hydrosylate.

* X A   Hydrogenated soybean oil – see soy, soy oil, hydrogenated vegetable oil; may be genetically modidied.

C A   Hydrogenated starch hydrolysate – 11; sugar alcohol; derived from corn, wheat or potato starch; may cause gastrointestinal distress; may be genetically modified; see sugar alcohols.

X A   Hydrogenated vegetable oil – associated

with heart disease, breast and colon cancer, atherosclerosis, elevated cholesterol; contains trans fats; may be derived from soy, corn, cottonseed or canola which are commonly GMO.

X A   <u>Hydrolyzed corn gluten</u> – contains MSG; may be genetically modified.

X A   <u>Hydrolyzed plant protein</u> – see hydrolyzed vegetable protein; may be genetically modified.

X A   <u>Hydrolyzed protein</u> – see hydrolyzed vegetable protein; may be GMO.

X A   <u>Hydrolyzed vegetable protein</u> – 22, 26; may cause brain and nervous system damage in infants; high salt content; may be corn, soy, or wheat based; contains free glutamates, MSG; may be genetically modified.

X A   <u>Hydrolyzed vegetable starch</u> – may contain MSG; may be GMO.

X A   <u>Hydroxyethyl cellulose</u> – 19, 48, 51; harmful if swallowed, inhaled and on skin contact, severe eye irritant, mutagenic.

X A   <u>Hydroxylated lecithin</u> – 7, 19; derived from soy lecithin, using microwave irradiation; soy may be GMO.

X A   <u>Hydroxypropyl cellulose</u> – toxic to lungs; eye, skin, gastrointestinal irritant; not adequately tested.

X A   <u>Hydroxypropyl methylcellulose</u> – eye, skin, respiratory, gastrointestinal irritant; not adequately tested.

C   <u>I3C</u> – see indole-3-carbinole.

X A   <u>Ice structuring protein</u> – made from genetically modified yeast containing a fish gene; may cause inflammation; used in ice cream products targeted to children; not adequately tested; long-term effects unknown.

X A   Imitation flavorings – see artificial flavoring.

X A   IMP – see disodium inosinate.

X A   Indigo carmine – see coal tar dyes.

C   Indole-3-carbinole – found in cruciferous vegetables; studies show contra-dictory results regarding cancer preven-tion; evidence that it may contribute to inflammatory diseases, heart problems, cancer when taken as a supplement.

* C   Inositol – a vitamin; see nutrient additives; may be genetically modified.

X   Interesterified fat – chemically modified fat; substitute for trans fats; may lower HDL (good cholesterol) even more than trans fats and raise blood sugar levels; may be derived from soy oil which is commonly genetically modified; not studied for long-term health effects.

* S   Inulin – naturally derived from chicory or made from sucrose; constituent of many herbs; dietary fiber; prebiotic; health promoting benefits for the intestinal tract; large amounts may cause flatulence or gastrointestinal distress.

* C A   Inversol – 49; see invert sugar.

* C A   Invert sugar – 49; half fructose, half glucose; usually derived from sucrose; promotes dental caries; see glucose and fructose; may be genetically modified.

C A   Inverse syrup – see invert sugar.

C A   Ionone – 9; caused swelling of liver cells in

75

	rats; eye, skin respiratory and digestive tract irritant; not adequately tested.
* C	<u>Iron, elemental</u> – see nutrient additives; least toxic form of iron used in foods.
* C	<u>Iron, reduced</u> – see nutrient additives, elemental iron.
C	<u>Iron oxide yellow</u> – may irritate skin, gastrointestinal tract; may be contaminated with crystalline silica, a carcinogen.
X	<u>Irradiated eggs</u> – deficient in vitamin A and niacin; disrupts protein-enzyme interacton in the egg; decreases nutritional value; whites are watery; approved by the FDA in 2001, but not proven safe; as of 3/07 not being used for commercial production.
X	<u>Irradiated iceberg lettuce</u> – approved in August 2008; see page 25.
X	<u>Irradiated Spices</u> – common if not organic; see page 25.
X	<u>Irradiated spinach</u> – approved in August 2008; see page 25.
C	<u>Isoamyl acetate</u> – 9; may cause headaches, fatigue, mucous membrane irritation.
* C A	<u>Isoascorbic acid</u> – 7; see erythorbic acid.
* X	<u>Iso-butane</u> – 41; neurotoxin; animal carcinogen according to NIOSH; may cause asphyxiation in large doses.
C	<u>Isoflavones</u> – may be genetically modified.
* X A	<u>Isolated soy protein</u> – 22; highly processed; may contain free glutamates; may be contaminated with nitrites; see free glutamates, soy isolates.
C	<u>Isomalt</u> – 4, 32, 49;excessive consumption may cause gastrointestinal distress; may be

	genetically modified. See sugar alcohols.
C	<u>Isomaltitol</u> – a combination of sugar and sugar alcohols; may cause gastrointestinal disturbances.
C	<u>Isomaltose</u> – a refined sugar.
X	<u>Isopentylamine</u> – 27; destructive to body tissues; toxic if swallowed; may be harmful to skin, eyes, lungs; not adequately tested.
* C A	<u>Isopropyl citrate</u> – 1, 7, 44; from citric acid and isopropyl alcohol; may interfere with results of medical lab tests.
X A	<u>ISP</u> – see ice structuring protein.
C	<u>Just Like Sugar</u>® - just like sugar, it will raise your blood sugar; best avoided.
* C A	<u>Karaya gum</u> – may cause gastrointestinal distress, dermatitis, asthma.
S	<u>Kelp</u> – from seaweed.
C	<u>Kola nut extract</u> – 9; contains caffeine.
C	<u>Lactitol</u> – see sugar alcohols.
S	<u>Lacto-fermented vegetables</u> – see fermented.
C	<u>L-ascorbic acid</u> – synthetic vitamin C; see nutrient additives; may cause diarrhea, kidney problems in large doses.
S A	<u>Lactalbumin</u> – 46, 50, 51; derived from milk.
C A	<u>Lactalbumin phosphate</u> – 46, 50, 51; see phosphates.
* C	<u>Lactase enzyme preparation from Kluyveromyces lactis</u> – see free glutamates.
* C A	<u>Lactic acid</u> – 1, 6, 7, 14, 27; may be corn or milk based; may be GMO.
* C	<u>Lactoflavin</u> – 10; vitamin B2; see nutrient additives; may be GMO.
C	<u>Lactose</u> – 37; may cause gastrointestinal distress.

C A	<u>Lactylic stearate</u>	– see calcium stearoyl lactylate.
C A	<u>Larch gum</u>	– see arabinogalactan.
X	<u>Laureate canola oil</u>	– genetically engineered rapeseed oil; see canola oil.
* C	<u>L-cysteine</u>	– amino acid; see nutrient additives.
* C	<u>L-cysteine monohydrochloride</u>	– amino acid; see nutrient additives.
C	<u>Leavening</u>	– may contain aluminum, BHA/BHT or chemical preservatives.
* C A	<u>Lecithin</u>	– 7, 17, 19, 46, 51; occurs naturally in egg yolks; most on market is from soy; watch for shift to sunflower-derived to avoid allergic effects of soy; may contain contaminants from solvent extraction and degumming; no studies showing beneficial effects on cholesterol or triglycerides; most is genetically modified; see soy lecithin.
C	<u>Leucine</u>	– an essential amino acid; may be genetically modified.
C A	<u>Levulose</u>	– see fructose.
* C	<u>Licorice</u>	- 27; extract of licorice root; safe for most people; when consumed in excess may cause high blood pressure, edema, sodium retention, loss of potassium, muscle and nervous system disorders; may have medicinal benefits for gastric ulcers.
* C	<u>Licorice extract</u>	– see licorice.
* C A	<u>Linoleic acid</u>	– may be corn, peanut, soy based; may be genetically modified.
X	<u>Lipase</u>	– enzyme derived from genetically modified fungi.
C	<u>Liquid fruit source</u>	– 49; 1 Tbsp. contains

		11 grams of sugars.
C A		<u>Litesse®</u> – fat substitute; see polydextrose.
C		<u>L-lysine</u> – amino acid; may be genetically modified; see nutrient additives.
C		<u>LMP 102</u> – bacteria-killing virus mixture to be sprayed on processed meat to prevent Listeria contamination; approved in 2006; unknown if it will be required to be listed on label.
S		<u>Lo Han</u> – 49; aka Lo Han Kuo and Lo Han Guo; used for centuries in China as a sweetener and medicinal herb; safe for hypoglycemics and diabetics; processed sweetener; use sparingly.
* S		<u>Locust bean gum</u> – see carob bean gum.
C A		<u>Lycasin</u> – see hydrogenated starch hydrosylate.
C		<u>Lycopene extract</u> – carotenoid derived from fruit used as natural coloring agent; solvent extracted using ethyl acetate.
* C		<u>Mace/nutmeg</u> – toxic to nervous system in large doses, can cause hallucinations; East Indian nutmeg may cause liver cancer.
C1		<u>Magnesium acetate</u> – 2, 4, 14, 39; magnesium is essential nutrient; may be harmful if kidney problems.
* C1		<u>Magnesium carbonate</u> – 2, 4, 13, 34; see nutrient additives; magnesium is essential nutrient; potentially harmful if kidney problems.
* C1		<u>Magnesium chloride</u> – 16, 23; see nutrient additives, magnesium carbonate.
C1A		<u>Magnesium fumarate</u> – 1; may be corn based; see magnesium acetate; may be genetically modified.

C1	<u>Magnesium gluconate</u> – see nutrient additives, magnesium carbonate.
* C1	<u>Magnesium hydroxide</u> – 2; see nutrient additives, magnesium carbonate.
* C1	<u>Magnesium oxide</u> – 39; see nutrient additives, magnesium carbonate.
* C1	<u>Magnesium phosphate</u> – see nutrient additives, magnesium carbonate.
* C1	<u>Magnesium silicate</u> – 4; skin, eye irritant; slightly hazardous if ingested, inhaled; not adequately tested; see magnesium acetate.
* X	<u>Magnesium stearate</u> – ingestion of large quantities is hazardous; skin, eye contact, inhalation slightly hazardous; may cause immune system suppression; may inhibit absorption of nutrients from intestinal tract; may be GMO; may be contaminated with pesticide residue; see nutrient additives, magnesium carbonate.
* C1	<u>Magnesium sulfate</u> – see nutrient additives, magnesium carbonate.
* C A	<u>Malic acid</u> – 1, 27; may be corn based; may be genetically modified.
C A	<u>Malt</u> – 27, 49; may contain barley, corn, wheat, or vinegar; see barley malt; may be genetically modified.
* C	<u>Malt extract</u> – 27, 49; see malt, barley malt; may be GMO.
C	<u>Malt flavoring</u> – may contain MSG.
* C	<u>Malt syrup</u> – 27, 49; see malt, barley malt; may be GMO.
C	<u>Malted barley</u> – see barley malt.
* C	<u>Maltitol</u> – 27, 32, 46, 49, 51; hydrogenated maltose; derived from corn; may cause gastrointestinal distress when consumed in

	large amounts; may be genetically modified.
C A	<u>Maltitol syrup</u> – see hydrogenated starch hydrosylate.
* X A	<u>Maltodextrin</u> – 26, 50; corn syrup solids; may be genetically modified; may contain free glutamates, MSG; see sucrose, MSG.
X	<u>Maltogenic amylase</u> – enzyme derived from genetically modified bacteria.
* C A	<u>Maltol</u> – 26; eye and skin irritant; may cause gastrointestinal, bladder, kidney, blood disturbances.
C A	<u>Maltose</u> – 49; may be corn, soy, wheat based; may be genetically modified.
X	<u>Maltrin® maltodextrins</u> – derived from hydrolyzed corn starch; may be GMO; may contain MSG.
* C	<u>Manganese chloride</u> – see nutrient additives; manganese is essential mineral; harmful in large amounts.
* C	<u>Manganese citrate</u> – see nutrient additives, manganese chloride.
* C	<u>Manganese gluconate</u> – see manganese chloride; see nutrient additives.
* C	<u>Manganese glycerophosphate</u> – see manganese chloride; see nutrient additives.
* C	<u>Manganese hypophosphite</u> – see nutrient additives, manganese chloride.
C	<u>Manganese oxide</u> – eye, skin, respiratory irritant; harmful if swallowed, inhaled; may adversely affect blood, kidneys, central nervous system; see nutrient additives.
* C	<u>Manganese sulfate</u> – see manganese chloride; see nutrient additives.
* C	<u>Manganous oxide</u> – see nutrient additives.
* C A	<u>Mannitol</u> – 32, 49, 50; small amounts may

	cause gastrointestinal distress; may cause nausea, kidney problems; sugar, seaweed or corn based; not thoroughly tested; may be genetically modified; see sugar alcohols
C	Maple syrup – 49; 1 Tbsp. contains 13 grams of sugars; contains healthful vitamins and minerals, most destroyed when heated; non-organic may be processed with formaldehyde, a carcinogen; use sparingly.
X A	Margarine – may contain genetically modified ingredients; see hydrogenated vegetable oil.
X A	Marmite – yeast extract; see MSG.
X	Maxarome Select – 26; "new tech-nology" yeast extract; intended to replace MSG; can be listed on label as yeast extract; contains free glutamates; company says all natural, does not define "new technology"; produced by DSM, one of largest biotech companies in Europe; most likely genetically engineered.
C	Menadione – vitamin K3; skin and mucous membrane irritant; see nutrient additives.
C	Menaquinone – vitamin K2; see nutrient additives.
* X	Menhaden oil – hydrogenated or par-tially hydrogenated; contains trans fats.
C	Methionine – amino acid; use as additive requires additional study; see nutrient additives.
* C	Menthol – nontoxic in concentrations less than 3%; irritant in concentrations above 3%; may cause nausea, vomiting, abdominal pain, coma in high concentrations.

* C A   Methylcellulose – 46, 51; may be derived from cotton; may be GMO.
* C A   Methyl ethyl cellulose – see carboxymethylcellulose.
* X   Methylparaben – 6, 40; synthetic preservative; mutagen; toxic if swallowed; associated with numerous health problems, including contact dermatitis and asthma; completely absorbed through skin; endocrine disrupter; may cause sensitivity in those sensitive to aspirin, or allergic to local anesthetics, like benzocaine, novocain; methyl-4-hydroxybenzoate is a natural paraben from blueberries which is safe; see parabens.
C   Methyl polysilicone – see dimethylpolysiloxane.
C   Methyl silicone – 5; see dimethylpolysiloxane.
C   Mica-Based Pearlescent Pigments – 38; formed from titanium salts and mica; may contain impurities of lead, arsenic, mercury;
C1   Microcrystalline cellulose – 4, 21, 24, 50; derived from vegetable, cereal or wood; no tests done regrding safety for infants and young children. Infants and young children should avoid.
C   Microcrystalline wax – 5; used in chewing gum; derived from crude oil; prolonged exposure may irritate skin; not adequately tested.
* C   Microparticulated protein product – made by heating in a mechanical rather than chemical process; fat substitute; not tested

for safety.
- C   <u>Milk powder</u> – may be GMO.
- C   <u>Milk solids</u> – in low-fat and no-fat milk; may contain MSG; see free glutamates, MSG.
- C   <u>Milo starch</u> – see modified food starch.
- * C   <u>Mixed carbohydrase and protease enzyme product</u> – may contain free glutamates; may be GMO.
- C   <u>Mixed carotenoids</u> – solvent extracted with harmful chemicals; may contain vegetable oil; see vegetable oil, nutrient additives.
- * X   <u>Modified cellulose gum</u> – see sodium carboxymethylcellulose.
- X A   <u>Modified corn starch</u> – see modified food starch.
- * X A   <u>Modified food starch</u> – 4, 32, 51; may be derived from corn; processed with chemicals of questionable safety; not adequately tested; may be GMO.
- X A   <u>Modified maltodextrin</u> – see modified food starch, maltodextrin.
- X   <u>Modified palm oil</u> – could be fractionated or hydrogenated; way of hiding hydrogenation from consumer.
- X A   <u>Modified starch</u> – see modified food starch.
- * S   <u>Molasses</u> – 49; 1 Tbsp. contains 9 grams of sugar; derived from sugar cane or sugar beets; contains iron, calcium, magnesium; use sparingly; see sucrose.
- * X A   <u>Mono- & diglycerides</u> – 5, 8, 17, 18, 19, 46; may be derived from animal fat, soy or canola oil, or be synthetically produced; always partially hydrogenated; contain trans fats; may be GMO; not adequately tested.

* X A   <u>Monoammonium glutamate</u> – 26; see ammonium, MSG.
* X   <u>Monobasic ammonium phosphate</u> – see ammonium, phosphates.
* C   <u>Monobasic calcium phosphate</u> – 44; see calcium phosphate.
* C   <u>Monobasic potassium phosphate</u> – see potassium phosphate.
* C   <u>Monocalcium phosphate</u> – see calcium phosphate.
* C A   <u>Monoglycerides</u> – see mono- & diglycerides.
* C A   <u>Monoisopropyl citrate</u> – 44; may interfere with results of medical lab tests; not adequately tested.
* X A   <u>Monopotassium glutamate</u> – 26; may cause nausea, gastrointestinal upset; see MSG.
* C   <u>Monopotassium phosphate</u> – see potassium phosphate.
* X A   <u>Monosodium glutamate</u> – 26; see MSG.
* C   <u>Monosodium phosphate</u> – 14, 19, 32; see phosphates.
* C   <u>Monosodium phosphate derivatives of mono- and diglycerides</u> – 19; see mono- and diglycerides.
* X A   <u>MSG</u> – 26; aka Accent, Aginomoto, Natural Meat Tenderizer; mutagen; causes obesity; addictive, makes you eat more; may cause diabetes, migraines, headaches, itching, nausea, brain, nervous system, reproductive disorders, high blood pressure, Autism, ADHD, Alzheimer's, retina damage, blindness; pregnant, lactating mothers, infants, small children should avoid; allergic reactions common; generally

produced using GMO bacteria; may be derived from corn; **may be hidden in** infant formula, baby food, low fat and no-fat milk, candy, chewing gum, drinks, kosher food, protein bars, protein powder, protein drinks recommended for seniors, most processed foods, wine, waxes applied to fresh fruits and vegetables, over-the-counter medications, especially children's, binders and fillers for nutritional supplements, prescription and non-prescription drugs, IV fluids given in hospitals, chicken pox vaccine, live virus vacines, nasal spray flu vaccine; used in pesticides, fungicides and fertilizers; being sprayed on growing fruits and vegetables as a growth enhancer (AuxiGro); proposed for use on organic crops. See free glutamates, processed free glutamates.

* C   Mustard oil - 27; see allyl isothio-cyanate.
* C A   Mycoprotein - meat substitute made from a mold-like fungus; may cause nausea, vomiting, diarrhea, cramps, hives, difficulty breathing, anaphylactic reactions; not adequately tested.
* C   N-butane - see isobutane.
* C   N-Butyric acid - 9; see butanoic acid.
   X   Natamycin - 6; antifungal drug used in cheese; no studies for use by pregnant, breast-feeding women, children, older adults.
   X   NatraTaste – see aspartame.
   X A   Natrium glutamate – contains MSG.
   X A   Natural beef flavoring – may contain MSG; see natural flavors.

X	A	<u>Natural chicken flavoring</u> – may contain MSG; see natural flavors.
X	A	<u>Natural flavorings</u> – see natural flavors.
X	A	<u>Natural flavors</u> – may be chemically extracted and processed and in combination with other food additives not required to be listed on the label; any ingredient whose purpose is to add flavor to a food rather than nutritional value; may contain free glutamates, MSG, may contain genetically modified ingredients; see MSG.
X	A	<u>Natural Meat Tenderizer</u> – see MSG.
X	A	<u>Natural pork flavoring</u> – may contain MSG; see natural flavors.
*	C	<u>Naturlose</u> – see tagatose.
*	X	<u>n-butane</u> – see butane.
	C	<u>nectresse</u> – 49; contains erythritol, sugar, molasses; see erythritol, sucrose.
	C	<u>Neohesperidin dihydrochalcone</u> – 11, 26; aka NHDC, Citrosa; may cause migraine headaches, nausea at 20 ppm.
	X	<u>Neotame</u> – 11, 26; aspartame + 3-dimethylbutyl; 3-dimethylbutyl is classified as hazardous by the EPA; approved for use in the U.S, 2002 and over 60 countries worldwide through 6/2010; ***used as a flavor enhancer or mixed with other sweeteners, it is NOT listed on the label***; industry sources cite at least 100 studies on the safety of neotame, however a search of PubMed and the National Library of Medicine could not find any published studies related to the safety of neotame; according to Jack Samuels, President of the Truth in Labeling Campaign, his wife requested a copy of

Monsanto's first application for approval of Neotame under the Freedom of Information Act. They were advised that they had to travel to Washington, D.C. to review the file, which they did. He reports, "At the time of our review of Monsanto's application, three human studies on the safety of Neotame were presented. The studies had few subjects, all of whom were employees of the company. Some of the subjects reported headaches after ingesting Neotame, but the researcher concluded that the headaches were not related to Neotame ingestion. Not mentioned in the studies was the fact that migraine headache is, by far, the most commonly reported adverse reaction to aspartame in the files of the FDA. The FDA has over 7,000 reports of adverse reactions to aspartame. The reported reactions include death"; not adequately tested.

* C   Neral – 9; see citral.

X   NET anthocyanin – nanotech color alternative to carmine; see NETColors.

X   NETColors – so called "natural colors" developed through nanotechnology; little is known about the effects of nanoparticles on living systems; nanotechnology ingredients have not been proven safe; studies with titanium dioxide have shown that nanoparticles can be absorbed into the cell and damage the DNA, can cross the blood-brain barrier; risks associated with nanoparticles are not adequately studied.

X   NET Tumeric – nanotech color; see

	NETColors.
* C	<u>Niacin</u> – vitamin B3; see nutrient additives; may cause flushing.
* C	<u>Niacinamide</u> – vitamin B3; see nutrient additives.
* X	<u>Nickel</u> – may cause dermatitis; ingestion of large amounts of nickel salts may cause kidney damage, gastrointestinal upset, nervous depression; carcinogen, IARC Group 1.
X	<u>Nickel sulfate</u> – nickel salt; carcinogen, IARC Group 1; see nickel.
C	<u>Nicotinamide</u> – see niacin; nutrient additives.
C	<u>Nicotinic acid</u> – see niacin; nutrient additives.
* C	<u>Nisin</u> – 40; bacteriocin, a gene encoded antimicrobial peptide or GMO, derived from lactic acid in pasteurized cheese; not required to be listed on the label.
X	<u>Nitrates</u> – 6, 16; form powerful cancer-causing agents in stomach; can cause death; considered dangerous by FDA but not banned because they prevent botulism; probable carcinogen, IARC Group 2A.
X	<u>Nitrites</u> – may cause headaches, nausea, vomiting, dizziness; probable carcinogen, IARC Group 2A; see nitrates.
* S	<u>Nitrogen</u> – 41.
C	<u>Nitrosyl chloride</u> – 13, 34; irritant to skin, eyes, mucous membranes; inhalation may cause pulmonary edema and hemorrhage.
* C1	<u>Nitrous oxide</u> – 41; has caused damage to fetus in pregnant lab animals.
* C	<u>Nori</u> – red algae; see brown algae.

89

* C A	<u>Nulomoline</u>	– 49; see invert sugar.
C	<u>Nutmeg</u>	– see mace/nutmeg.
X	<u>Nutrasweet</u>	– 11; ake aspartame, Equal, Spoonful; see aspartame.
X	<u>Nutra Taste</u>	– see aspartame.
C	<u>Nutrient additives</u>	– nutrients added to mostly refined and processed foods giving a false sense of nutritional value and can lead to nutritional imbalances; too much or too little of individual nutrients can lead to imbalance of other nutrients in the body; chemicals used in preparing nutrients added are not listed on the label; may be derived from natural or synthetic sources; most nutrients added to processed foods are synthetic, highly processed and derived from petroleum and processed with toxic chemicals, like formaldehyde; may be genetically modified.
* C A	<u>Oat gum</u>	– 7, 46, 51; may cause gastrointestinal distress.
C	<u>Oil of caraway</u>	–9; see d-carvone.
C	<u>Oil of cloves</u>	– 27; may cause gastrointestinal irritation in lab animals.
C	<u>Oil of Mace</u>	– see mace/nutmeg.
C	<u>Oil of Nutmeg</u>	– see mace/nutmeg.
* C	<u>Oil of rue</u>	– 27; may cause photo-sensitivity; not adequately studied for carcinogenic effects.
X A	<u>Okara</u>	– soy pulp, by-product of soy milk and tofu production; insoluble fiber; contributes to chronic digestive problems; may be GMO.
X	<u>Olean</u>	– see olestra.
* C	<u>Oleic acid</u>	– 5, 12, 24; skin irritant, low oral

	toxicity; may be GMO.
X	Olestra – 21; causes gastrointestinal irritation, reduces carotenoids and fat soluble vitamins in the body.
* S	Oligofructose – may be naturally derived from chicory or synthesized from sucrose; dietary fiber; prebiotic; health promoting benefits for the intestinal tract; large amounts may cause flatulence or gastrointestinal distress.
X	Orange B – coal tar dye; restricted; may cause exposure to cancer-causing agent.
* S	Ox bile extract – see bile salts.
* C A	Oxystearin – 4, 5, 17, 19; may be soy based; may be genetically engineered.
C A	PABA – see para aminobenzoic acid.
C	Palm kernel oil – most commercial palm kernel oil is refined, deodorized, bleached and used in processed foods; it is not only unhealthy, but the farming producing these oils is contributing to the destruction of the rainforests; unrefined, non-hydrogenated, non-bleached, non-deodorized palm kernel oil is safe and healthy.
C	Palm oil – most commercial palm oil is refined, deodorized, bleached and used in processed foods; it is not only unhealthy, but the farming producing these oils is contributing to the destruction of the rainforests; unrefined, non-hydrogenated, non-bleached, non-deodorized palm oil is safe and healthy.
C	Pantothenic acid – vitamin B5; see nutrient additives.
* C A	Papain – 15, 35; skin, eye irritant; may

cause stomach upset; meat tenderizer; protein-digesting enzyme from papaya; Hawaiian papaya may be genetically modified; has caused birth defects in lab animals.

* C Paprika, paprika oleoresin – 38; may be harmful in large amounts.

C A Para aminobenzoic acid – may cause light sensitivity, rash, swelling when applied to skin in sensitive individuals; insufficient data to evaluate human carcinogenicity, IARC Group 3.

X A Parabens – 6, 40; may be toxic if swallowed; potential mutagen; endocrine disrupter; may impair fertility; in recent studies, parabens have been found in breast cancer tumors, but it is unknown if they had a part in causing the tumors; not adequately tested; do not use on children.

X Parasobic acid – causes indigestion, may cause kidney damage; converted to sorbic acid when heated under pressure; causes cancer in rats, no human data available; IARC, Group 3; not adequately tested.

X A Partially hydrogenated vegetable oil – see hydrogenated vegetable oil.

X Pasteurization with x-rays – irradiated.

C Pasteurized eggs – heated to kill Salmonella without cooking the egg; all egg products, i.e. eggs not in the shell, are required by law to be pasteurized; significantly different in appearance and performance than unpasteurized eggs, whites are cloudy, watery, yolks are pale; detectable difference in taste; denatures some of the protein.

X A	<u>Pea protein</u>	– hydrolyzed protein; contains high levels of glutamic acid, MSG; see free glutamates, MSG, hydrolyzed vegetable protein.
* C A	<u>Peanut oil</u>	– cold pressed peanut oil may contain traces of peanut protein and is unsafe for those with peanut allergies; highly refined peanut oil does not contain traces of peanut protein and is safe for those with peanut allergies, however, highly refined oils are not healthy oil choices; those with peanut allergies should choose cold pressed, non-peanut oils.
C	<u>Pear oil</u>	– see amyl acetate.
X	<u>Pectinesterase</u>	– enzyme derived from genetically modified fungi.
* C	<u>Pectins</u>	– 19, 46, 51; commercially may be prepared with ammonia; may contain free glutamates, MSG; may cause flatulence, bloating; see MSG.
X	<u>PEG</u>	– see polyethylene glycol.
C	<u>Pepsin</u>	– protein-digesting enzyme secreted by stomach; may be GMO.
* C	<u>Peptones</u>	– 46; partially hydrolyzed protein; may contain MSG.
C	<u>PGPR</u>	– see polyglycerol polyricinoleic acid.
C	<u>Phenylalanine</u>	– avoid if phenyl-ketonuric; component of aspartame; listing this on the label may be a way of hiding aspartame; may be genetically modified; see aspartame.
X	<u>Phenylmethyl cyclosiloxane</u>	– 4; has caused liver, kidney, reproductive damage in lab animals.
X	<u>Phenylpropanolamine</u>	– causes headaches,

		increased blood pressure, strokes and death; women 18-49 at risk of hemorrhagic stroke.
C		Phosphates – 1, 19, 44, 50; can inhibit mineral absorption; excess consumption can cause kidney damage, osteoporosis.
C		Phosphatidyl serine – generally derived from soy; scientific studies showing benefits use bovine-derived, containing DHA, which is absent in soy.
* X		Phosphoric acid – 1, 44; extreme skin, eye, nose, throat and respiratory irritant; corrosive; skin contact and swallowing are highly toxic; extremely environ-mentally toxic; see phosphates.
C		Phylloquinone – vitamin K1; see nutrient additives.
C A		Phytic acid – may be GMO.
C		Phytonadione – vitamin K1; see nutrient additives.
X		Pickled vegetables – possible carcino-gen, IARC Group 2B; as prepared traditionally in Asia; some studies show association with stomach or esophageal cancer.
X		Pinang – carcinogen, IARC Group 1.
* C		Piperonal – 9; chemical used to kill lice.
C		Plant sterol esters – plant sterols are found naturally in plant foods; the esters are used primarily in margarines and processed foods; infants, children, pregnant and lactating women should avoid; those using cholesterol medication should seek advice from their doctor before using; may cause gastrointestinal discomfort; not studied for effects of long-term usage.
C		Poloxamer 335 – derived from petroleum;

indirect food additive; may be toxic to respiratory, immune systems; see page 139.

C A   Polydextrose – 11, 32, 50; fat substitute; may be corn based; may cause gastro-intestinal distress; laxative effect; may be GMO.

X   Polyethylene glycol – skin and eye irritant; slightly toxic when ingested; potential mutagen; may be contaminated with carcinogen 1,4-dioxane; can break down into formaldehyde.

C   Polyglycerol polyricinoleic acid – 19; fatty acid ester of castor oil used mostly in chocolate; inadequate data on safety available.

X A   Polyoxyethylene-40-monostearate – 5, 19; may be contaminated with 1,4-dioxane; not adequately tested.

X A   Polyoxyethylene stearate – 50; may cause gastrointestinal upset, skin rashes, kidney problems; may be contaminated with 1,4-dioxane; chemically modified mono- & diglycerides.

C A   Polysorbate 60 – 19; may be contaminated with carcinogen, 1,4-dioxane.

C A   Polysorbate 80 – see polysorbate 60.

C A   Polysorbates – 5, 19; sorbitol and corn, peanuts or soy based; possible contamination with carcinogen, 1,4-dioxane; may be derived from genetically engineered ingredients.

X A   Ponceau 4R – coal tar dye; carcinogen.

C   Potassium acetate – may cause kidney problems.

* C1   Potassium acid tartrate – see cream of

        tartar.

* C   Potassium alginate – 46, 51; avoid if pregnant, kidney or heart problems; see alginates.

* C A   Potassium benzoate – combines with vitamin C (ascorbic acid) to form cancer-causing benzene; may cause skin rashes, gastrointestinal upset; avoid if kidney, liver, heart problems.

* X1   Potassium bicarbonate – 33; see nutrient additives; avoid if kidney or heart problems; causes cancer in rats.

  X A   Potassium bisulfite – 6; not permitted in meats, foods with Vitamin B1, raw fruits and vegetables; no studies done to evaluate if carcinogenic to humans, IARC Group 3; see sodium bisulfite.

  X   Potassium bromate – 34; can cause nervous system, kidney disorders, gastrointestinal upset; causes cancer in animals; possible carcinogen, IARC Group 2B; banned worldwide except Japan and the U.S.

* C1   Potassium carbonate – 2; see nutrient additives; skin irritant; avoid if kidney or heart problems.

  C A   Potassium caseinate – 50; avoid if kidney or heart problems; see caseinates.

* C   Potassium chloride – 43; see nutrient additives; can cause nausea, gastrointestinal irritation; caution if kidney, liver or heart problems.

* C A   Potassium citrate – 1, 7, 14, 44; see nutrient additives; may cause mouth ulcers; avoid if kidney or heart problems.

  C1A   Potassium fumarate – 1; may be corn

                based; avoid if kidney or heart problems; may be GMO.

* C1A   Potassium gluconate – 14, 23, 26, 44; see nutrient additives; may be corn based; avoid if kidney or heart problems; may be GMO.
* X A   Potassium glutamate – 26; see MSG.
* C   Potassium glycerophosphate – see nutrient additives.
* X   Potassium hydroxide – may cause mouth ulcers, gastrointestinal upset; severe skin and eye irritant; corrosive; has caused tumors on skin of mice; ingestion can be fatal; aka potash or potash lye.
* C1   Potassium iodate – 4, 18; caution if kidney, heart or thyroid problems; should be used under care of physician.
* C1   Potassium iodide – see potassium iodate.
* C A   Potassium lactate – 27; see lactic acid.
  X A   Potassium metabisulfite – 6; not permitted in meats, foods with Vitamin B1, raw fruits and vegetables; no studies done to evaluate if carcinogenic to humans, IARC Group 3; see sodium bisulfite.
  X   Potassium nitrate – 6, 16; see nitrates.
  X   Potassium nitrite – 6, 16; see nitrites.
* C   Potassium phosphate – 52; avoid if kidney, heart problems; see phosphates.
  C   Potassium polyphosphate – may cause gastrointestinal upset; avoid if kidney or heart problems.
  X A   Potassium prosulfite – no studies done to evaluate if carcinogenic to humans, IARC Group 3; see sodium bisulfite.
* C   Potassium sorbate – 6, 40; can be naturally occurring, but most is made synthetically;

skin, respiratory, digestive irritant; see sorbic acid.

* C   Potassium sulfate – harmful if inhaled, swallowed, absorbed through skin; may cause mouth ulcers, burning sensation in mouth, gastrointestinal disturbances; large amounts can cause gastrointestinal bleeding; toxological properties not fully investigated.

X A   Potassium sulfite – 3, 7, 40; no studies done to evaluate if carcinogenic to humans, IARC Group 3; see sodium bisulfite.

X   PPA – see phenylpropanolamine.

X A   Processed free glutamic acid – glutamic acid that has been manufactured and contains both d-glutamic and l-glutamic acid. L-glutamic acid, as found in protein and the human body, is bound to other amino acids in long chains and causes no adverse reactions. D-glutamic acid occurs as a result of the manufac-turing process. It is a neurotoxin and causes brain lesions and neuroendocrine disorders in lab animals. In humans, it causes a great many symptoms, including skin rashes, tachycardia, migraine headaches, depression, and seizures. See MSG, free glutamates.

* X   Propane – 41; may be narcotic, neurotoxin in high doses.

X   1,2-Propanediol – see propylene glycol.

* C A   Propane-1,2,3-triol – see glycerol.

* C A   1,2,3-Propanetriol – see glycerol.

* C A   Propionic acid – 6; may trigger behavior changes.

* X A   Propyl gallate – 7; slightly toxic on

ingestion; associated with dermatitis, anemia, kidney, liver problems, gastrointestinal irritation; considered non-carcinogenic by National Toxicology Program even though it caused cancer in male rats; not adequately tested.

* X A Propylene glycol – 24, 32, 50; "antifreeze;" skin and eye irritant, has caused nervous system and kidney damage in animals; ingestion can cause blindness, kidney failure, convulsions and death.

* X A Propylene glycol alginate – 5, 46, 51; avoid if pregnant; see propylene glycol, alginates.

* X Propylene glycol monostearate – 18, 19; may be corn, peanut, soy based; combined with hydrogenated vegetable oil; may be genetically modified.

X Propylene oxide – possible carcinogen, IARC Group 2B; on the EPA Extremely Hazardous Substances list.

* X A Propylparaben – 6, 40; skin irritant; associated with numerous of health problems, including contact dermatitis and asthma; endocrine disrupter; strong allergen; research shows it decreases sperm production; propyl-4-hydroxybenzoate is a natural paraben which is safe; see parabens.

C Protease – enzyme for protein digestion; occurs naturally in the body; also an exotoxin that can cause cellular damage; may contain MSG; may be genetically modified from bacteria.

C Protease enzymes – see protease; may contain MSG.

C Protein concentrates – may contain MSG;

	may be genetically modified.
C	<u>Protein fortified</u> – may contain MSG; may be genetically modified.
C	<u>Protein isolates</u> – may contain MSG; may be genetically modified.
C	<u>Protein powder</u> – may contain MSG; may be genetically modified.
C	<u>Protein supplements</u> – may contain MSG; may be genetically modified.
X	<u>Pullulanase</u> – enzyme derived form genetically modified bacteria.
S	<u>PuraQ Xtend</u> – natural antimicrobial made from polylysine, by natural fermentation.
C	<u>PurVia</u> – 49; made from stevia extract, dextrose, cellulose powder and natural flavors; may contain GMOs, hidden MSG.
* C	<u>Pyridoxine hydrochloride</u> – vitamin B6; see nutrient additives.
C	<u>Quercetin</u> – a flavonoid, naturally occurring in fruits and vegetables; anti-inflammatory; antioxidant; anti-histamine; anti-tumor properties; beneficial dose not known; long-term adverse effects of excess unknown; may enhance effect of chemotherapy drugs; best to get from foods; supplements should be taken under guidance of health care practitioner; has caused cancer in rats.
C	<u>Quercetin dihydrate</u> – did not show carcinogenic effects on rats; see quercetin.
* C	<u>Quicklime</u> – 1, 18, 52; see calcium oxide.
X A	<u>Quinine</u> – 27; may impair hearing, cause birth defects; very poorly tested.
C A	<u>Quorn</u> – see mycoprotein.
S	<u>Raftiline</u> – see inulin.

S   Raftilose – see oligofructose.

S   Rapadura – organic, whole cane sugar; a whole food containing vitamins and minerals; use sparingly.

* X   Rapeseed oil – 19, 46; contains high erucic acid, associated with fibrotic heart lesions when there is a selenium deficiency; fresh, unprocessed rapeseed oil used in China, Japan and India in ancient times caused no adverse effects with adequate saturated fats in diet; genetically manipulated in 1970's to produce low euricic acid rapeseed oil, renamed canola oil; the problem is modern processing at high temperatures, with toxic solvents, refining, bleaching, degumming and deodorizing; deodori-zing transforms omega-3 fatty acids into trans fatty acids; see canola oil.

X   rBGH – recombinant bovine growth hormone, injected in U.S. dairy cattle; rBGH milk and non-rBGH milk may be mixed in 80-90% of U.S. milk, includ-ing infant formula; banned in all Indus-trialized countries except U.S. because it harms animal health; causes increased levels of IGF-1 in milk; increased levels of IGF-1 in human blood correlated with obesity, and cancers of the breast, prostate, colon (Amer. Cancer Society, 1998); has caused cancer in rats (Canadian government scientists, 1998); genetically engineered; see References for "Who offers rBGH-Free Milk?".

X   rBST – recombinant bovine somatotrophin; see rBGH.

	C	<u>Reb A</u> – extracted from the stevia leaf; may be extracted with solvents or alcohols.
	C	<u>Rebaudioside A</u> – derived from stevia; see Reb A.
	C	<u>Rebiana</u> – derived from Reb A; see Reb A.
*	C	<u>Red algae</u> – see brown algae.
	X A	<u>Red No. 3</u> – 10; see FD&C Red No. 3.
	X A	<u>Red No. 40</u> – 10; see FD&C Red No. 40.
*	C1A	<u>Reduced lactose whey</u> – avoid if milk allergies.
*	C1A	<u>Reduced minerals whey</u> – avoid if milk allergies.
*	C	<u>Rennet</u> – 20; enzyme for making cheese; can be animal rennet from calf stomach (derived from rennin), vegetable rennet from plants like fig leaf or thistle, microbial rennet from cultured fungiform microorganisms or genetically modified (GMO) rennet, a genetically engineered form of animal rennet; avoid GMO rennet; commercially prepared rennet may contain propylene glycol or sodium benzoate as preservatives.
	C	<u>Resolver® natural flavors</u> – designed to block bitter tastebuds to make food taste better; new and proprietary technology; company says non-GMO, non-nanotechnology, non-irradiated, but would not answer any questions about how the ingredients are processed.
	C	<u>Ribitol</u> – see sugar alcohols.
*	C	<u>Riboflavin</u> – 10; vitamin B2; see nutrient additives; may be GMO.
*	C	<u>Riboflavin-5-phosphate</u> – vitamin B2; see nutrient additives.

S		Rice amasake – 49; 1 Tbsp. contains 2 grams of sugars.
C		Rice syrup – 49; see barley malt.
C		Rice syrup powder – 49; 1 Tbsp. contains 4 grams of sugars; has some minerals and complex carbohydrates; all sweeteners are best avoided; see sucrose.
C		RiceLife – hydrolyzed whole grain brown rice; hydrolysis process may cause the formation of MSG.
* C		Rue – see oil of rue.
X		Saccharin – 11; delisted as a carcinogen in 1997, however, studies still show that saccharin causes cancer; IARC Group 3; see sodium saccharin.
X		Safflower oil – very high in omega-6 fattyacids, which are already too high in Western diets; many store bought oils are processed at high heat, causing the formation of trans fats.
* C		Saffron – 38; may be harmful in large amounts; not adequately tested.
C1		Sage – avoid if pregnant, epileptic or high blood pressure.
* C A		St. John's bread gum – locust bean gum.
X		Salatrim – 21; aka Benefat™; inter-esterified fat; may contain hydrogenated canola, soy, cottonseed or sunflower oil; may be genetically modified; may cause headaches, flatulence, diarrhea, nausea, stomach cramps; not adequately tested.
C A		Salicylates – avoid if aspirin sensitive.
* C		Salt – may be genetically modified if contains dextrose; see sodium chloride.
X		Salted fish (Chinese-style) – carcinogen,

	IARC Group 1.
* C	<u>SAPP</u> – see sodium acid pyrophosphate.
C	<u>Seasonings</u> – as an ingredient, may contain MSG; may be irradiated.
C	<u>Selenium</u> – trace mineral essential for health in small amounts; toxic if taken in excess; IARC Group 3; has caused mammary tumors in female mice.
* X	<u>SelenoExcell®</u> – high selenium yeast; GMO; flagship product of biotech company, Cypress Systems.
X	<u>Senomyx</u> – biotech company that produces flavors that enhance taste; do not appear on the label; listed under artificial or natural flavors; widely used, but companies refuse to identify products containing these genetically engineered flavors.
X	<u>Shortening</u> – generally refers to hydrogenated vegetable oil; contains trans fats.
C	<u>Shoyu</u> – may be genetically modified.
X	<u>Shugr™</u> - 49; blend of tagatose, erythritol, maltodextrin and sucralose.
C	<u>Silica</u> – 4; may be associated with kidney problems; occupational inhalation of crystalline silica is carcinogenic, IARC Group 1; occupational inhalation of amorphous silica not sufficient data to determine carcinogenicity, IARC Group 3.
* C	<u>Silica aerogel</u> – 4; see silica.
C	<u>Silicates</u> – 4; see silica.
* C	<u>Silicon dioxide</u> – 4; see silica.
C	<u>Silicones</u> – 5; may be associated with kidney problems.
X	<u>Silver Nitrate</u> – antimicrobial in bottled

        water; small amounts used FDA considers safe; poison if taken internally; hazardous on skin contact, ingestion; may damage skin, eyes, mucous membranes; severe overexposure can cause death.

* C A   Simplesse – 21; fat substitute; "microparticulated whey protein concentrate" made from egg white and milk protein; heated in a mechanical rather than chemical process; made by the makers of Nutra-Sweet; received GRAS status without being tested for safety.

X   Smoked flavoring – smoked foods may contain nitrosamines and polycyclic aromatic hydrocarbons (PAH) formed during the smoking process; PAH's and nitrosamines are carcinogenic; smoked flavoring may reduce exposure; not adequately tested.

X   Smoked yeast – see smoked flavoring; not adequately tested; questionable safety.

* C   Sodium acetate – 14, 40; eye, skin, respiratory tract irritant; large amounts may cause gsstrointestinal distress; may cause blood pressure, kidney disturb-ances, water retention.

* C   Sodium acid phosphate –44; see sodium acid pyrophosphate.

* C   Sodium acid pyrophosphate – 14, 33; skin, eye and respiratory irritant; can inhibit mineral absorption; may cause blood pressure, kidney disturbances, water retention; not adequately tested.

* C   Sodium alginate – 17, 46, 51; may cause

* C		blood pressure, kidney disturbances, pregnancy complications; see alginates.
* C		Sodium aluminum phosphate – see aluminum, sodium acid pyrophosphate.
C		Sodium aluminum sulfate – 13, 34; causes kidney failure in lab animals; see aluminum, sodium acetate.
* C		Sodium aluminosilicate – 4; may be associated with kidney problems, mineral malabsorption.
* C A		Sodium ascorbate – 7; see nutrient additives; synthetic vitamin C; may be corn based; may contribute to blood pressure, kidney disturbances, water retention; may be genetically modified.
* X A		Sodium benzoate – 6, 40; combines with vitamin C (ascorbic acid) to form cancer-causing benzene; skin irritant; toxic if swallowed; avoid if asthma or liver problems; may cause hyperactivity in children; associated with numerous health issues; insufficient data to support safety in products where exposure involves inhalation. see benzoate of soda.
* S		Sodium bicarbonate – 14, 33; baking soda; relatively safe; avoid consuming large amounts; large amounts may cause gastrointestinal distress.
X A		Sodium bisulfite – 3, 6, 7; destroys vitamin B1; small amounts may cause asthma, anaphylactic shock; dangerous for asthma, allergy sufferers; has caused deaths; banned on fresh fruits and vegetables, except potatoes; no studies done to evaluate if carcinogenic to humans, IARC Group 3.

X		Sodium calcium alginate – 19, 23, 46, 51; safety not determined.
* C		Sodium calcium aluminosilicate – 4; may be associated with kidney problems, mineral malabsorption.
* C A		Sodium carboxymethylcellulose – 4, 17, 51; see carboxymethylcellulose.
* C		Sodium carbonate – 39; may cause heart problems, gastrointestinal distress; see sodium acetate.
* X A		Sodium caseinate –50; avoid if milk allergy or MSG sensitive; contains free glutamates, MSG; see MSG, casein.
* C		Sodium chloride – 26, 40; may cause blood pressure, kidney disturbances, water retention.
* C		Sodium citrate – 1, 7, 14, 19, 27, 44; see sodium acetate, citric acid.
X A		Sodium cocoyl glutamate – see glutamates.
C		Sodium copper chlorophyllin – slightly hazardous on skin or eye contact; not adequately tested.
* C		Sodium diacetate – 6; see sodium acetate.
* C A		Sodium erythorbate – 7; may be corn based; see sodium ascorbate.
X		Sodium ferrocyanide – 4; extremely hazardous if ingested; severe skin and eye irritant; toxic to mucous membranes, blood, lungs.
X		Sodium fluoride – rat poison; causes dental fluorosis, mottling of the teeth; may cause premature aging, arthritis, bone cancer in boys, calcium loss, joint stiffness, brittle bones, weight loss, weakness, anemia; toxic to heart, liver, kidneys; ingestion may cause

death; toxic effects may be delayed; ADA advises that babies under one year of age should NOT be given fluoridated water; IARC Group 3.

- C   <u>Sodium formate</u> – may cause urinary tract problems.
- C A   <u>Sodium fumarate</u> – 1; may be corn based; see sodium ascorbate; may be genetically modified.
- * C A   <u>Sodium gluconate</u> – 44; see sodium ascorbate.
- * C   <u>Sodium hexametaphosphate</u> – 19, 44, 46, 51; see sodium acid pyrophosphate.
- X A   <u>Sodium hydrosulfite</u> – no studies done to evaluate if carcinogenic to humans, IARC Group 3; see sodium bisulfite.
- * X   <u>Sodium hydroxide</u> – 2, 39; caustic soda, lye; alters protein content of food; severe eye and skin irritant; corrosive; mutagen.
- * C A   <u>Sodium isoascorbate</u> – see sodium ascorbate.
- * C   <u>Sodium lactate</u> – 7, 23, 32; see sodium acetate.
- X A   <u>Sodium lauryl sulfate</u> – skin, eye, mucous membrane irritant; skin contact may cause allergic reactions; moderately toxic if ingested; may cause mutagenic effects.
- X A   <u>Sodium metabisulfite</u> – no studies done to evaluate if carcinogenic to humans, IARC Group 3; see sodium bisulfite.
- * C   <u>Sodium metaphosphate</u> – 18, 44; see sodium acid pyrophosphate.
- * C   <u>Sodium metasilicate</u> – see silica, sodium acid pyrophosphate.
- X   <u>Sodium nitrate</u> – see nitrates.

X	<u>Sodium nitrite</u> – see nitrites.
* C	<u>Sodium pantothenate</u> – see nutrient additives, sodium acetate.
* C	<u>Sodium pectinate</u> – 46, 51; see sodium acetate.
* C	<u>Sodium phosphate</u> – 44; see sodium acid pyrophosphate.
C	<u>Sodium polyphosphate</u> – may cause gastrointestinal upset; see sodium acid pyrophosphate.
* C	<u>Sodium potassium tartrate</u> – 1, 19; may cause blood pressure, kidney disturbances; avoid if heart problems.
* C A	<u>Sodium propionate</u> – 6, 40; may trigger headaches, behavioral changes, blood pressure, kidney disturbances, water retention.
* C	<u>Sodium pyrophosphate</u> – 44; see sodium acid pyrophosphate.
S	<u>Sodium riboflavin phosphate</u> – vitamin B2; see nutrient additives.
X	<u>Sodium saccharin</u> – causes cancer in rats; see saccharin.
X	<u>Sodium selenite</u> – form of selenium often found in nutritional supplements; on the EPA Extremely Hazardous Substances list; selenium is essential in small amounts, see your healthcare provider to determine your individual needs; see selenium.
* C	<u>Sodium sesquicarbonate</u> – 39; see sodium acetate.
* C	<u>Sodium silicate</u> – 4; see sodium acetate.
C	<u>Sodium silicoaluminate</u> – 4; see sodium calcium aluminosilicate.
* C	<u>Sodium sorbate</u> – 3, 6; may cause blood

pressure, kidney, liver disturbances, water retention.

C   <u>Sodium starch glycolate</u> – eye, skin, inhalation irritant; not adequately tested.

C A   <u>Sodium stearoyl fumarate</u> – 18, 19; may be corn, milk, peanut, soy based; see sodium acetate, stearic acid; may be genetically modified.

C A   <u>Sodium stearoyl lactylate</u> – 18, 19, 46; see sodium stearoyl fumarate.

C   <u>Sodium sulfate</u> – 14; may cause heart, kidney problems; see sodium acetate.

X A   <u>Sodium sulfite</u> – 3, 6, 40; no studies done to evaluate if carcinogenic to humans, IARC Group 3; see sodium bisulfite.

C A   <u>Sodium sulfoacetate mono- & diglycerides</u> – 19; not adequately tested.

* C   <u>Sodium tartrate</u> – 1, 19; see sodium acetate.

* C   <u>Sodium thiosulfate</u> – 7; see sodium acetate.

* C   <u>Sodium tripolyphosphate</u> – 40, 44, 46, 50, 51; can cause calcium depletion; skin and mucous membrane irritant; may cause violent vomiting or diarrhea if ingested; GRAS for packaging.

C   <u>Softeners</u> – see ester gum.

* C   <u>Sorbic acid</u> – 3,6, 40; may be drived from berries (from parasorbic acid heated under pressure); most is chemically synthesized; may cause skin rashes, abdominal pain; interferes with enzyme functions in the body; high doses injected under the skin has caused cancer in lab animals; mildly toxic when ingested; possible mutagen; low oral toxicity; see parasorbic acid.

C A   <u>Sorbitan monopalmitate</u> – 19; see

	polysorbates.
C A	<u>Sorbitan monostearate</u> – 5, 19, 46; see polysorbates, stearic acid.
C A	<u>Sorbitan tristearate</u> – 19; see polysorbates, stearic acid.
* C A	<u>Sorbitol</u> – 11; may cause extreme gastrointestinal distress, bloating, diarrhea, abdominal pain, especially in infants and children; DO NOT give to infants and children; may be corn based; may be genetically modified; see sugar alcohols.
C A	<u>Sorbitol syrup</u> – see hydrogenated starch hydrosylates.
C	<u>Sorghum molasses</u> – 49; see sucrose.
S	<u>Sorghum syrup</u> – 49; see barley malt.
X	<u>Soy</u> – contains toxins that cannot be completely removed with processing, enzyme inhibitors that block enzymes needed for digestion, phytates that inhibit mineral absorption; promotes clumping of red blood cells, kidney stones; depresses thyroid function; weakens immune system; fermented soy products have less toxins; commonly genetically modified (89%).
X A	<u>Soy concentrates</u> – 22; may contain MSG; commonly genetically modified; see soy isolates, free glutamates, MSG.
X	<u>Soy flour</u> – may be GMO; see soy.
X	<u>Soy isoflavones</u> – potent endocrine disruptors; may cause growth problems, thyroid disorders, kidney stones, weak immune system, food allergies, increased breast cancer risk, leukemia, colon cancer, infertility, decreased sperm production; mutagenic; teratogenic; most soy is

genetically modified.

* X A   <u>Soy isolates</u> – 22; may inhibit nutrient absorption; may be contaminated with nitrites; may contain MSG; commonly genetically modified; see soy, free glutamates, MSG.

C A   <u>Soy lecithin</u> – from soybean waste; may be contaminated with solvent and pesticide residues; contains trace amounts of soy protein; should NOT be used as a nutritional supplement; may cause the formation of nitrosamines from choline metabolism in the intestines; may be genetically modified; see soy.

X A   <u>Soy milk</u> – see soy; commonly genetically modified.

X A   <u>Soy oil</u> – contaminated with hexane from chemical extraction at high temperatures; commonly GMO.

X A   <u>Soy protein</u> – hydrolyzed protein; may contain MSG; commonly GMO; see soy, MSG, free glutamates, hydrolyzed vegetable protein.

X A   <u>Soy protein concentrate</u> – see soy concentrates.

X A   <u>Soy protein isolate</u> – see soy isolates.

X A   <u>Soy sauce</u> – may contain MSG: see soy; MSG, free glutamates; may be genetically modified.

X A   <u>Soy sauce extract</u> – may contain MSG: may be GMO; see soy; MSG, free glutamates.

C A   <u>Spice</u> – generic term to protect trade secrets; may be a combination of many different spices; may be fumigated or irradiated.

X	<u>Splenda</u> – sucralose + maltodextrin + dextrose; maltodextrin and dextrose derived from high fructose corn syrup; 99% pure sugar; may adversely affect blood sugar; laws allow to say sugar-free if less than .5 g sugar per serving and calorie free if less than 5 calories per serving; may be GMO; see sucralose; not tested for safety.
X	<u>Spoonful</u> – see aspartame.
* X	<u>Stannous chloride</u> – 7; eye, skin, mucous membrane, respiratory, gastrointestinal irritant; corrosive; ingestion may cause diarrhea, nausea, vomiting, abdominal pain, stomach bleeding, reduction of blood pressure, convulsions; may cause kidney, liver damage; possible mutagen, teratogen.
* X A	<u>Starch</u> – 51; may be corn, wheat based; may be genetically modified; see modified food starch.
* C	<u>Starter distillate</u> – 9; see diacetyl.
* X	<u>Stearic acid</u> – 5, 12; suppresses immune system; may be derived from hydrogenated oils; may contain high leves of pesticide residue; may be genetically modified.
* X	<u>Stearidonic Acid (SDA) Omega-3 Soybean Oil</u> – soybean oil genetically engineered to have a high omega-3 content.
C A	<u>Stearoyl lactylate</u> – 18, 19; see calcium stearoyl lactylate, sodium stearoyl lactylate.
X A	<u>Stearoyl propylene glycol hydrogen succinate</u> – 19; breaks down into propylene glycol and fatty acids; see propylene glycol, stearic acid.
C A	<u>Stearoyls</u> – 18, 19; may be corn, milk,

           peanut, soy based; see stearic acid; may be genetically modified.

* C A   Stearyl citrate – 7, 44; may be corn based; may be genetically modified; see stearic acid, calcium citrate.

   C   Stellar – 21; not adequately tested.

* C   Sterculia gum – see vegetable gum.

   C   Sterol esters – naturally occurring in plant foods and oils; shown to lower cholesterol; may be derived soy; most soy in U.S. is genetically modified.

* S   Stevia – 49; has been safely used for hundreds of years, without adverse reactions in South America, as a sweetener and for its therapeutic value; has been used in Japan as a food additive for over 30 years with no reports of adverse reactions; makes up over 40% of Japanese sweetener market; has been used in China for over 17 years with no reports of adverse reaction; whole leaf stevia is rich in vitamins and minerals; there is conjecture that the anti-stevia camp is trying to protect the artificial sweetener market; can be used by those with candida, diabetes, hypoglycemia.

* S   Steviol – steviol glycosides are aka stevioside.

* S   Stevioside – sweet component of stevia; does not contain the vitamins and minerals of whole leaf stevia, but is anti-bacterial and anti-viral.

   X   Stock – may contain MSG.

* C   STPP – see sodium tripolyphosphate.

   X   Substance 951 – GMO ingredient that enhances sweetness of sugar; see Senomyx.

C	A	<u>Sucanat</u> - 49; 1 Tbsp. contains 12 grams of sugars; refined sweetener; same as sugar with a few minerals; all sweeteners are best avoided; use sparingly; see sucrose.
* X		<u>Succinic acid</u> - 14, 27, 39; corrosive; harmful if ingested, absorbed through skin or inhaled; may cause gastro-intestinal disturbances, laxative effect; not adequately tested.
X	A	<u>Succistearin</u> - 19; see stearoyl propylene glycol hydrogen succinate.
X		<u>Sucralite</u> – see sucralose
X		<u>Sucralose</u> – 11; chlorinated sugar; ; chlorocarbon, aka organochlorine, i.e. insecticide; made from chemically treated sucrose; chlorine is a human carcinogen according to the EPA; may break down into toxic gases when heated and chlorosugars when stored for several months; weakly mutagenic; contrary to manufacturer's claims, sucralose is partially absorbed and metabolized by the body, 11-27% according to the FDA and 40% according to the Japanese Food Sanitation Council; may cause skin rashes, eruptions, hives, respiratory distress, headaches, migraines, inflammation or swelling of face, lips, tongue, throat, eye irritation, digestive disturbances, chest pain, heart palpitations, joint pain, seizures, dizziness, depression, mood swings, numbness; may cause adverse effects in diabetics; may be stored in liver, kidneys, gastrointestinal tract; contains small amounts of dangerous contaminants, such as heavy metals, methanol, arsenic,

triphenilphos-phine oxide, chlorinated disaccharides, chlorinated monosaccharides; a 2008 study on male rats showed a 50% reduction of beneficial flora in the intestines, increased intestinal pH, increased body weight and altered factors which reduce your body's ability to use orally administered drugs; only five small human studies done on middle-aged men as of 2006, longest was three months; no studies done on children or pregnant women; no monitoring of adverse health effects; those with chlorine allergies may suffer severe reactions; not adequately tested.

C Sucroglycerides – 19; may be produced in presence of toxic solvents.

* C Sucrose – 49; primarily derived from sugarcane or sugar beets; associated with blood sugar problems, depression, fatigue, B-vitamin deficiency, hyperactivity, tooth decay, periodontal disease, indigestion; may be genetically modified if from sugar beets; action being sought to ban GMO sugar beets; as of August 2010, Monsanto cannot plant any more GE sugar beets until they provide an environmental impact statement, expected in 2012, but the biotech industry may try to rush it through sooner; GE sugar beets already on the market may have found their way into the food chain.

C Sucrose polyester – 21; not adequately tested.

C Sugar – see sucrose; may be derived from genetically engineered sugar beets.

C	<u>Sugar alcohols</u> – synthesized by hydrogenating sugars; mannitol, sorbitol, xylitol, lactitol, isomalt, maltitol, erythritol,, glycerol and hydrogenated starch hydrolysates; may cause stomach cramps, laxative effect, bloating, diarrhea, flatulence; may cause carbohydrate cravings; may contribute to candida, yeast problems; may cause metabolic acidosis, acid reflux, increased risk of cancer of larynx; raise blood sugar levels; promote dehydration, loss of electrolytes; exercising after consuming may increase risk of muscle cramps, heat stroke, cardiovascular problems; may increase frequency of seizures in epileptics.
X	<u>Sulfites</u> – IARC Group 3; see sodium bisulfite.
* X	<u>Sulfur dioxide</u> – 3, 7, 40; ; inadequate evidence to classify as a carcinogen, IARC Group 3; see sodium bisulfite; on the EPA Extremely Hazardous Substances list.
X	<u>Sweet 'n Low</u> – 11; contains saccharin.
* X	<u>Talc</u> – carcinogen if contaminated with asbestos, IARC Group 1; free of asbestos, IARC Group 3, for inhalation.
* S	<u>Talin</u>® - 26, 49; see thaumatin.
* C	<u>Tagatose</u> – 11; aka D-Tagatose, Naturlose; sugar alcohol; may cause increased blood uric acid levels and enlarged liver; may cause nausea, flatulence, diarrhea.
* S	<u>Tallow</u> – stable fat; does not usually become rancid; no free radicals form with normal use; foods fried in tallow usually have less fat than foods fried in liquid vegetable oils;

	contains anti-microbial palmitoleic acid.
X	Tamari – may be genetically modified; may contain MSG; see soy, MSG, free glutamates.
* X	Tannic acid – 15, 27; carcinogenic in rats, no data on humans, IARC Group 3.
* X	Tannin – see tannic acid.
C A	Tara gum – 46, 51; indigestible; may cause gastrointestinal disturbances if consumed in large quantities.
* S	Tartaric acid – 1; may cause gastro-intestinal distress.
X A	Tartrazine – 10; FD&C Yellow No. 5; may cause breathing difficulty, hay fever, skin rashes, blurred vision; avoid if aspirin sensitive.
X A	Taste No. 5 – see MSG.
C	Taurine – amino acid produced by the body; produced synthetically in the lab for energy drinks and other energy products; see nutrient additives.
X	TBHQ – 7; skin, eye, respiratory irritant; harmful if swallowed, inhaled, absorbed through skin; not adequately tested.
C	Tempeh – one of three soy products with least amount of soy toxins; may contain MSG; may be genetically modified; see MSG, soy.
X	Tertiary butylhydroquinone – see TBHQ.
X	Tetrachloroethylene – used in food packaging; probable carcinogen, IARC Group 2A.
* X	Tetrasodium pyrophosphate – 19, 44; poison if ingested; can cause nausea, vomiting, diarrhea; GRAS for packaging.

X A		Textured protein – see textured vegetable protein.
X A		Textured soy protein (TSP) – see textured vegetable protein.
X A		Textured vegetable protein (TVP) – 22; made from soy; aka textured soy protein (TSP); chemically processed at very high heat; may be contaminated with carcinogenic and mutagenic hetero-cyclic amines; may be contaminated with aluminum; contains high levels of phytates which bind minerals and make them unavailable to your body; contains free glutamates, MSG; see MSG; most soy is genetically modified.
* S		Thaumatin – 26, 49; sweet protein derived from West African Katemfe berries; 3000 times sweeter than sucrose; technology exists to create this protein through genetic engineering, but not utilized as of 7/2010.
X		THBP – 7; questionable safety; not adequately tested.
C		Theobromine – extracted from cacao bean, used to synthesize caffeine; no data on carcinogenicity, IARC Group 3
* C		Thiamine hydrochloride – synthetic vitamin B1; see nutrient additives.
* C		Thiamine mononitrate – synthetic vitamin B1; see nutrient additives.
* S		Thiodipropionic acid – 7, 40.
C		Threonine – essential amino acid; may be genetically modified.
* S A		Thyme oil – 27.
X		Ticagel® 795 Powder – see carrageenan.
X		Ticagel® 822 – see carrageenan.

C		Ticagel® Bind-KX EC – may contain carrageenan.
ϕ X		Titanium dioxide – 38; may irritate skin; inhalation of large amounts of titanium dioxide dust may cause lung damage; use limited to 1% by weight; possibly carcinogenic to humans, IARC Group 2B, associated with inhalation and with fine and ultrafine particles.
X		Titanium dioxide (micronized) – possibly carcinogenic to humans, IARC Group 2B; see titanium dioxide.
* C A		Tocopherol – 7; vitamin E; see nutrient additives; may be corn, peanut, soy based; may be genetically modified.
X		Tofu – may be GMO; see soy.
* C A		Tragacanth – 46, 51; can cause gastro-intestinal distress; not adequately tested.
C		Trailblazer® – 21; fat substitute; see microparticulated protein product.
X		Trans fat – mostly from partially hydrogenated vegetable oils; associated with heart disease, breast and colon cancer, atherosclerosis, elevated cholesterol; may increask for infertility; food labels may say "no trans fat" if it contains less than 0.5 grams per serving; no amount of trans fat is safe.
C		Treacle – 49; may contain corn; may be genetically modified; see corn syrup, invert sugar.
* C		Trehalose – 46, 49; derived from corn; may cause gastro-intestinal distress when consumed in large amounts; may be genetically modified.

* C		Triacetin – see glyceryl triacetate.
* C		Tribasic calcium phosphate – see calcium phosphate.
* C		Tribasic potassium phosphate – see potassium phosphate.
* C		Tricalcium phosphate – see calcium phosphate.
* C		Tricalcium silicate – 4; see silicates.
* S A		Triethyl citrate – 44; may interfere with medical lab test results.
X		2-4-5 Trihydroxybutrophenone – see THBP.
* C A		1,2,3-Trihydroxypropane – see glycerol.
* C A		Tripotassium citrate – see potassium citrate.
* C		Trisodium phosphate – 44, 50; see sodium acid pyrophosphate.
C		Truvia – 49; made from erythritol, rebiana, natural flavors; may contain GMOs, hidden MSG.
X A		TSP – see textured soy protein.
* X		TSPP – see tetrasodium pyrophosphate.
* C		Tumeric – 38; may be extracted with harmful solvents.
C		Turbinado sugar – 49 ; 1 Tbsp. contains 12 grams of sugars; refined sweetener; all sweeteners are best avoided; use sparingly; see sucrose.
* C		Turmeric – same as tumeric.
X A		TVP – see textured vegetable protein.
X		Ultra-pasteurized – may contain MSG.
X A		Umami – see MSG.
C		Urea – eye, skin and respiratory irritant; ingestion causes gastrointestinal irritation, diarrhea, nausea, vomiting; may cause electrolyte depletion, headache, confusion;

	aka carbamide, isourea, pseudourea, carbonyl diamine, carbaderm, Supercel 3000.
C	<u>Vanilla extract</u> – may be from corn; may be genetically modified.
* X	<u>Vanillin</u> – 9; see ethyl vanillin.
C	<u>Vegapure® Sterol Esters</u> – may be derived from soy, corn, rapeseed or sunflower oil; most soy and corn in U.S. is genetically modified; see sterol esters.
X A	<u>Vegemite</u> – yeast extract; see MSG
C	<u>Vegetable broth</u> – 27; may contain corn, additives not listed on the label; may be a source of hidden MSG; see MSG, free glutamates; may contain GMO ingredients.
X	<u>Vegetable fat</u> – may be derived from corn, soy, canola; may be hydrogenated; may be genetically modified.
C A	<u>Vegetable gum</u> – 19, 46, 51; may be derived from corn; not adequately tested.
X	<u>Vegetable oil</u> – may be partially hydrogenated; may be derived from soy, corn, cottonseed or canola which are commonly genetically modified.
C	<u>Vegetable oil sterols</u> – see plant sterol esters.
X	<u>Vegetable protein</u> – may contain MSG; may be genetically modified; see MSG, free glutamates, soy.
X	<u>Vegetable shortening</u> – hydrogenated oil; associated with cardiovascular disease; contains trans fats; may be genetically modified.
X	<u>Veratraldehyde</u> – 26; see ethyl vanillin.
X A	<u>Vetsin</u> – see MSG.

* C   <u>Vitamin A</u> – dietary supplement; may be toxic in very large doses; see nutrient additives.

* C   <u>Vitamin A acetate</u> – synthetic vitamin A; may be toxic in very large doses; see nutrient additives.

* C   <u>Vitamin A palmitate</u> – synthetic vitamin A; may be toxic in very large doses; see nutrient additives.

* C   <u>Vitamin B1</u> – dietary supplement; see nutrient additives.

* C   <u>Vitamin B2</u> – dietary supplement; see nutrient additives; may be GMO.

* C   <u>Vitamin B3</u> – dietary supplement; see nutrient additives.

* C   <u>Vitamin B6</u> – dietary supplement; see nutrient additives.

* C   <u>Vitamin B12</u> – dietary supplement; may be GMO; see nutrient additives.

* C   <u>Vitamin C</u> – dietary supplement; see nutrient additives; may be genetically modified if made in North America.

* C   <u>Vitamin D2</u> – synthetic vitamin D; may be toxic in very large doses; see nutrient additives.

* C   <u>Vitamin D3</u> – natural vitamin D; may be toxic in very large doses; see nutrient additives.

  C   <u>Vitamin E</u> – see tocopherol, nutrient additives; may be genetically modified.

  C   <u>Vitamin G</u> – vitamin B2; see nutrient additives.

  C   <u>Vitamin K</u> – may be listed on label also as Vitamin K_1, Phylloquinone, Menaquinone-4, Vitamin K_2, Menadione, Menadione sodium bisulfite, Menadione sodium bisulfite trihydrate, Menadiol, Menadiol sodium phosphate, Menadiol sodium phosphate hexahydrate; Acetomenaphthone; Phylloquinones (include K and K_1) derived from food and rarely exhibit toxic effects; Menadiones synthetic, derived from petroleum, may be associated with cellular damage and haemolytic anemia; neither has been investigated for mutagenic effects.

S   Vivapigments™ - natural plant derived colors.

* S   Wellmune wgp® - beta glucan derived from yeast; non-GMO.

S A   Wheat bran

S A   Wheat germ

* C1A Wheat gluten – 18; avoid if gluten sensitive or malabsorption syndrome.

X A   Wheat protein – hydrolyzed protein; contains MSG; see free glutamates, MSG, hydrolyzed vegetable protein.

* C1A Whey – 12; avoid if lactose intolerance or milk allergies; may be GMO.

C A   Whey powder – may be GMO.

C A   Whey protein – may be hydrolyzed; may contain MSG; see free glutamates, MSG, hydrolyzed protein; may be GMO

* C A Whey protein concentrate – may contain free glutamates, MSG; may be GMO; see soy isolates, MSG, whey.

* X A Whey protein hydrolysate – see hydrolyzed protein.

* C A Whey protein isolate – may contain free glutamates, MSG; may be GMO; see soy isolates, MSG, whey.

* C A	White thyme oil	– 27; may cause gastrointestinal upset, dizziness, cardiac depression.
C	White vinegar	– may be GMO.
* C1	Wild cherry bark	– therapeutic benefit; should not be used by pregnant and nursing women, those with low blood pressure; may cause sedation; not to be taken for extended periods of time; leaves and pits contain poison.
C	Wild Resolver® natural flavors	– see Resolver® natural flavors.
X	Wormwood	– 27; see artemisia.
C A	Xanthan gum	– 48, 51; may be derived from corn; may cause gastrointestinal distress; extracted from Xanthomonas campestris by solvent extraction which may leave a toxic residue; may contain allowable amounts of lead, arsenic and heavy metals; may be GMO.
C	Xylitol	– 11; synthetic; highly processed industrial product; no studies done on safe levels for children; amounts greater than 30 grams/day for adults may cause gastrointestinal distress, laxative effect; may cause increased uric acid levels, liver problems; has caused cancer in lab animals at high doses; claims that xylitol prevents tooth decay are not fully supported; NOT safe for pets; "xylitol poisoning in dogs causes weakness, loss of consciousness and seizures"; may be derived from corn which is commonly GMO; see sugar alcohols.
X A	Yeast autolyzates	– contains free glutamates, MSG; not adequately tested; see

	MSG.
X A	<u>Yeast extract</u> – 26; contains free glutamates, MSG; see MSG.
X A	<u>Yeast food</u> – contains free glutamates, MSG; see MSG.
X A	<u>Yeast-malt sprout extract</u> – 26; contains free Glutamates, MSG; see MSG.
* S A	<u>Yellow beeswax</u> – 19, 27, 29.
X A	<u>Yellow No. 5</u> – 10; see FD&C Yellow No. 5
X A	<u>Yellow No. 6</u> – 10; see FD&C Yellow No. 6
X	<u>Yellow prussiate of soda</u> – 4; see sodium ferrocyanide.
* C A	<u>Zein</u> – corn protein; may be GMO.
* C	<u>Zinc chloride</u> – see nutrient additives; excess can cause anemia, gastro-intestinal distress, mild skin irritation.
* C	<u>Zinc gluconate</u> – see nutrient additives, zinc chloride.
C	<u>Zinc methionine sulfate</u> – see nutrient additives, zinc chloride.
* C	<u>Zinc oxide</u> – see nutrient additives, zinc chloride; may be unsafe if micronized.
* C	<u>Zinc stearate</u> – see nutrient additives, zinc chloride, stearic acid.
* C	<u>Zinc sulfate</u> – see nutrient additives; excess can cause anemia, gastro-intestinal distress, mild skin irritation; has caused tumors in lab animals.

OTHER FOOD ADDITIVES ...
YOU WON'T FIND ON THE LABEL

There are three kinds of additives not included in the ingredients list on food packages.

Ingredients in the Ingredients

Manufacturers sometimes include ingredients in the ingredient list that contain other ingredients. Examples are broth, spices, natural flavors and seasonings.

The ingredients contained in the ingredient listed are not required to be noted on the label. In other words, the ingredients in broth will ***not*** appear in the ingredients list. This is the way manufacturers hide food additives.

The ingredients mentioned above are all hiding places for MSG. In fact, ***packages may state "No MSG" or "No MSG Added" and still contain these hidden sources of MSG!*** You will find the ingredients that are hiding places for MSG under the additive "Free glutamates" on page 70.

These and other food additives that contain other ingredients will be noted in the main Food Additives listing, starting on page 28.

Secondary Direct Food Additives

These are Ingredients used in the manufacturing and processing of food additives. They are processing aids, used for "a technical effect in food during processing." It is *claimed* they are "not in the finished food." However, allowable amounts of residue are permitted. For example…

- ***acetone***, a petroleum derivative, is an extremely toxic chemical which can cause kidney, liver and nerve damage. It is used in the extraction of spices, and is allowed to leave a residue of 30 parts per million.
- ***ethylene dichloride*** may cause cancer, kidney and liver damage and nervous system depression. It is an irritant to the digestive system and may cause nausea, vomiting and diarrhea. Used in spice extraction, a residue of 30 parts per million is allowed.
- ***isopropyl alcohol*** may cause digestive symptoms, such as cramps, nausea, diarrhea, vomiting, or more severe symptoms like unconsciousness and death. It is permitted to have a 6 parts per million residue in the production of lemon oil and 50 parts per million from spice extraction.
- ***hexane*** may cause nervous system damage. and adverse gastrointestinal symptoms such as nausea, vomiting and diarrhea. A residue of 25 parts per million is allowed in spice extraction.
- ***trichloroethylene*** is toxic to the nervous system, the liver, kidneys, heart and upper respiratory tract. It is hazardous if ingested. Allowable residue varies from 10 parts per million for decaffeinated coffee to 30 parts per million for spices.

These and other secondary additives may be avoided or minimized by choosing organic brands that use non toxic, solvent-free processes.

Food Contact Substances

The third type of additive you won't find listed on the label is Food Contact Substances. By definition, they are "any substance intended for use as a component of materials used in manufacturing, packing, packaging, transporting, or holding food if such use is not intended to have a technical effect in such food." Also known as **indirect additives**, food contact substances are chemicals in the paper, plastic and cellophane used to package your food that are not directly added to your food, but that may be transferred to the food items upon contact with the food.

The presence of these undisclosed chemicals that may contaminate the food you're buying is not only in commercial, non-organic packaged foods, but also in organic packaged foods. This is a very good reason to minimize your purchases of packaged and processed foods and choose primarily fresh, whole, organic foods.

FOODGARD™, developed by a Canadian R&D company, introduced in 2011, is a "line of Certified Organic, functional ingredients for the purpose of reducing microbial loads in food processing applications." It is an organic alternative to food processing chemicals.

You cannot tell by the package what kind of indirect additives, i.e. food contact substances, have come in contact with the packaged food you buy.

REFERENCES

A. Branen, P. Davidson, S. Salminen, Food Additives. New York: Marcel Dekker, Inc., 1990.

A Broken Record How the FDA Legalized – and Continues to Legalize – Food Irradiation Without Testing It for Safety, http://www.citizen.org/documents/brokenrecordfinal.PDF

Aflatoxins, http://vm.cfsan.fda.gov/~mow/chap41.html

Russell L. Blaylock, M.D., Excitotoxins: The Taste That Kills. Snata Fe, NM: Health Press, 1997.

Botanical.com, http://botanical.com/botanical/mgmh/mgmh.html

Carcinogenic Potency Project, http://potency.berkeley.edu/chemnameindex.html

Center for Science in the Public Interest, http://www.cspinet.org/reports/chemcuisine.htm. http://www.cspinet.org/reports/saccomnt.htm

CFR - Code of Federal Regulations Title 21, http://www.accessdata.fda.gov/scripts/cdrh/cfdocs/cfcfr/CFRSearch.cfm?fr=73.350

Compendium of Food Additive Specifications. Addendum 5. (FAO Food and Nutrition Paper - 52 Add. 5), http://www.fao.org/docrep/W6355E/w6355e00.HTM

Crystalline Fructose, http://thegoldenspiral.org/2009/01/27/crystalline-fructose/

Kaayla, T. Daniel, Ph.D., CCN, The Whole Soy Story: The Dark Side of America's Favorite Health Food.

Washington, D.C.: New Trends Publishing, Inc, 2005.

Does Your Supplement Contain this Potentially Hazardous Ingredient?, http://articles.mercola.com/sites/articles/archive/2012/06/23/whole-food-supplement-dangers.aspx?e_cid=20120623_DNL_artNew_1

Mary G. Enig, Ph.D., Know Your Fats, Silver Spring, MD: Bethesda Press, 2004.

EPA, Agriculture, http://www.epa.gov/oecaagct/nahs.html

EPA Extremely Hazardous Substances (EHS) http://www.ehs.columbia.edu/EPAWasteCode.html

John E. Erb & T. Michelle Erg, The Slow Poisoning of America. Paladins Press, 2003.

FDA Legalized Food Irradiation Without Adequately Evaluating Its Risks, Dismissed Evidence of Serious Public Health Hazards, http://www.preventcancer.com/press/releases/oct3_00.htm

FDA OKs using viruses to fight *Listeria* in meat, http://www.cidrap.umn.edu/cidrap/content/fs/food-disease/news/aug2206phage2.html

David B. Fankhauser, Genetically Engineered Fish Protein In Ice Cream: A Good Idea?, 2006, http://biology.clc.uc.edu/fankhauser/Society/GMO/GM_foods_ice_cream.htm

Fluoride - "an emerging neurotoxin," The Lancet 2007, http://www.fannz.org.nz/lancet.php

Food Intolerance Network Factsheet: Annatto (160b), http://fedup.com.au/factsheets/additive-and-natural-chemical-factsheets/160b-annatto

Food Irradiation: FDA Could Improve Its Documentation and Communication of Key Decisions on Food Irradiation Petitions, http://www.gao.gov/assets/100/96545.pdf

Nicholas Freydberg, Ph.D. and Willis A. Gortner, Ph.D., The Food Additives Book. New York: Bantam Books, 1982.

Gary Gibbs, Deadly Dining: The Food That Would Last Forever. http://www.mindfully.org/Food/Food-Last-Forever-TOC.htm

Ann Louise Gittleman, Get The Sugar Out: 501 Simple Ways To Cut The Sugar Out Of Any Diet. New York: Crown Trade Paperbacks, 1996.

Glycerol ester of tall oil rosin, http://www.fao.org/ag/agn/jecfa-additives/specs/monograph7/additive-516-m7.pdf

Glycerol ester of tall oil rosin, http://www.fao.org/ag/agn/jecfa-additives/specs/monograph11/additive-516-m11.pdf

Mark Gold, Aspartame/NutraSweet Toxicity Info Center, http://www.holisticmed.com/aspartame

Nikki & David Goldbeck, The Goldbeck's Guide To Good Food. New York: New American Library, 1987.

Robert Goodman, A Quick Guide To Food Safety. San Diego: Silvercat Publications, 1992.

Susan Gordon, Editor, Critical Perspect Ives On Genetically Modified Crops And Food. New York: The Rosen Publishing Group, Inc., 2006.

Health Alert Newsletter.

Honey, GMO Compass, http://www.gmo-compass.org/eng/database/food/238.honey.html

How U.S. FDA's GRAS Notification Program Works, http://www.fda.gov/Food/FoodIngredientsPackaging/GenerallyRecognizedasSafeGRAS/ucm083022.htm

Robert S. Igoe, MS, MBA and Y. H. Hui, Ph.D., Dictionary of Food Ingredients, 4th edition. Gaithersburg, MD: Aspen Publishers, Inc., 2001.

INCHEM, International Programme on Chemical Safety, http://www.inchem.org/

International Agency for Research on Cancer, http://monographs.iarc.fr/ENG/Classification/index.php

International Federation for Produce Standards, http://plucodes.com/docs/IFPS-plu_codes_users_guide.pdf

Irradiated Foods, http://www.cdc.gov/ncidod/dbmd/diseaseinfo/foodirradiation.htm#whichfoods

Grace Ross Lewis, 1001 Chemicals In Everyday Products. John Wiley & Sons, 1999.

Linus Pauling Institute, Micronutrient Information Center, http://lpi.oregonstate.edu/infocenter/
LYCOPENE EXTRACT FROM TOMATO, http://www.fao.org/fileadmin/templates/agns/pdf/jecfa/cta/71/lycopene_extract_from_tomato.pdf

Material Safety Data Sheets (MSDS), http://www.ilpi.com/msds/

MedlinePlus, http://www.nlm.nih.gov/medlineplus

Dr. Joseph Mercola, http://www.mercola.com

Joseph Mercola, Sweet Deception: Why Splenda®, Nutrasweet®, and the FDA May Be Hazardous to Your Health. Nashville, TN: Nelson Books, 2006.

Earl Mindell, Unsafe At Any Meal. New York: Warner Books, 1987.

Bill Misner, Ph.D., Xylitol research, Director of Research & Product Development E-Caps Inc. & Hammer Nutrition Limited

National Fire Protection Association (NFPA) Chemical Hazard Ratings, http://www.delta.edu/slime/nfpakey.html

National Toxicology Program (NTP) Report on Carcinogens, http://ntp.niehs.nih.gov/index.cfm?objectid=72016262-BDB7-CEBA-FA60E922B18C2540

Kathy R. Niness, Inulin and Oligofructose: What Are They?, Journal of Nutrition. 1999;129:1402S-1406S.

NIOSH Chemical Databases, http://www.cdc.gov/niosh/database.html

Phytosterols, Oregon State University, Linus Pauling Institute, http://lpi.oregonstate.edu/infocenter/phytochemicals/sterols/index.html

Phytosterols, PDRhealth, http://www.pdrhealth.com/drug_info/nmdrugprofiles/nutsupdrugs/phy_0205.shtml

Michael Pollan, "Playing God in the Garden: New York Times Magazine on GE Crops," The New York Times Sunday Magazine, October 25, 1998.
http://www.organicconsumers.org/ge/playinggd.htm

Dr. Pollen, "Pet Poison: Xylitol,"
http://drpollen.blogspot.com/2006/09/pet-poison-xylitol.html

Ranking Possible Cancer Hazards from Rodent Carcinogens, Using the Human Exposure/Rodent Potency Index (HERP), http://potency.berkeley.edu/pdfs/herp.pdf

Jack L. Samuels, Truth in Labeling Campaign, http://www.truthinlabeling.org

Doris Sarjeant & Karen Evans, HARD TO SWALLOW: The Truth About Food Additives. Canada: Alive Books, 1999.

Scientific Opinion on the Safety of Glycerol Esters of Tall Oil Rosin for the proposed uses as a food additive, http://www.efsa.europa.eu/en/efsajournal/pub/2141.htm

Scorecard, http://scorecard.goodguide.com/

Scorecard, Good Guide, Extremely Hazardous substances, http://scorecard.goodguide.com/chemical-groups/one-list.tcl?short_list_name=ehs

Second Opinion Newsletter.

Gilles-Eric Séralini, et. al., Long term toxicity of a Roundup herbicide and a Roundup-tolerant genetically modified maize. Food and Chemical Toxicology, Volume 50, Issue 11, November 2012, Pages 4221–4231, http://www.sciencedirect.com/science/article/pii/S0278691512005637

Jeffrey M. Smith, Genetic Roulette: The Documented Health Risks of Genetically Engineered Foods. Yes! Books, 2007, www.GeneticRoulette.com

Jeffrey M. Smith, SEEDS OF DECEPTION: Exposing Industry and Government Lies about the Safety of the Genetically Engineered Foods You're Eating. White River Junction, VT: Chelsea Green Publishing, 2003.

Stearidonic (SDA) Omega-3 Soybean Oil, http://www.accessdata.fda.gov/scripts/fcn/gras_notices/grn 000283.pdf

Keith Steinkraus, Editor, Handbook of Indigenous Fermented Foods. New York: Marcel Dekker, Inc., 1996.

David Steinman & Samuel S. Epstein, M.D., THE SAFE SHOPPER'S BIBLE. New York: MacMillan, 1995.

Stephen Stiteler, L.Ac., O.M.D., "Coffee – It's Time to Avoid It!" http://www.drstephenstiteler.com/hidden-dangers-coffee/

The Mystery Behind Organic Honey, http://livingmaxwell.com/organic-honey-certified

The Top 10 Problems With Irradiated Food, http://www.soc.iastate.edu/sapp/pctop10.pdf

The Weston A. Price Foundation, http://www.westonaprice.org

ToxFAQs™, http://www.atsdr.cdc.gov/toxfaqs/index.asp

Truth About Splenda, http://www.truthaboutsplenda.com

U.S. Food and Drug Administration, COLOR ADDITIVE STATUS LIST, http://www.fda.gov/ForIndustry/ColorAdditives/ColorAdditiveInventories/ucm106626.htm

Melanie Warner, Food Companies Test Flavorings That Can Mimic Sugar, Salt or MSG. April 6, 2005.

http://www.nytimes.com/2005/04/06/business/06senomyx.html?ex=1270440000&en=05d1bc83162499e3&ei=5090&partner=rssuserland

WHO Food Additives Series 46:D-Tagatose, http://www.inchem.org/documents/jecfa/jecmono/v46je04.htm#_46042200

Who offers rBGH-Free Milk?, Organic Consumers Association, http://www.organicconsumers.org/rBGH/rbghlist.cfm

Ruth Winter, A CONSUMER'S DICTIONARY OF FOOD ADDITIVES. New York: Crown Publishers, 1999.

Ruth Winter, POISONS IN YOUR FOOD. New York: Crown Publishers, 1991.

At KISS For Health, our mission is to help you be safe from harmful ingredients by showing you how to be an informed consumer so you know what's safe to use and what's not.

We also search the marketplace to find products that have the safest and healthiest ingredients.

Information on KISS For Health books can be found on the web at:

www.kiss4healthpublishing.com
www.healthyeatingadvisor.com/bookstore.html
www.dyingtolookgood.com

The Healthy Eating Advisor has an abundance of healthy eating information and links to healthy products and other healthy websites at:

www.healthyeatingadvisor.com

DyingToLookGood.com offers a wealth of information and resources for protecting yourself from harmful chemicals in your cosmetics and personal care products at:

www.dyingtolookgood.com

The following websites offer healthy products, and services recommended by Dr. Farlow:

www.healthyeatingadvisor.com/customnutrition.html
www.dyingtolookgood.com/shop.html
dyingtolookgood.miessence.com/home.jsf
www.christinehfarlowdc.com
drfarlow.mybeyondorganic.com

Books By Dr. Farlow

DYING TO LOOK GOOD: The Disturbing Truth About What's Really in Your Cosmetics, Toiletries and Personal Care Products... And What You Can Do About It

FOOD ADDITIVES: A Shopper's Guide To What's Safe & What's Not

Available through:

KISS For Health Publishing
PO Box 462335-F13
Escondido, CA 92046-2335
(760) 735-8101
e-mail: kiss4health@lycos.com

Order by mail – see page 140.
Telephone orders – pay by phone check – call (760) 735-8101.

Payment must be in U.S. funds drawn on a U.S. bank only.

Order online at
www.foodadditivesbook.com
www.healthyeatingadvisor.com/bookstore.html
www.amazon.com
www.dyingtolookgood.com

Order Form

Qty	Title	Price
	DYING TO LOOK GOOD (2006 edition) $12.95 ea	
	FOOD ADDITIVES (2013 edition) $8.95 ea	
S&H	U.S. $5.95 Canada $12.95 International $15.95 S&H rate for 1-4 books.	
Subtotal		
Tax (California residents 8.00%)		
Total		

Send check or money order (U.S. funds drawn on a U.S. bank only) to:

**KISS For Health Publishing
PO Box 462335-F13
Escondido, CA 92046-2335**

To pay by phone check, for contact information or to order online, see page 139.

This little book is packed with helpful information. It guides, educates, and gives you the heads up when it comes to shopping. I keep a copy in my glove box in my car to grab when I go to the grocery store. I think that everyone who eats needs to know the dangers of food additives--they're in virtually everything we eat! This little book will help you assess the risks and make informed decisions on your purchases.
Julian

Excellent, just the right size for purse. Great reference book. It comes in handy when checking the ingredients panel. Helpful and easy to read.
DHill

I wish I knew about this book before now! Had I known about this book earlier, I would have made more responsible choices. In reality, I've switched to organic food and chosen several alternatives that are much better for me. I refer to this book and share my findings with my friends and family. The book is easy to understand and the author has codes to determine what is generally recognized as safe, and additives with an X signifies that it is unsafe. *Grgygirl*

Great little book to keep in my bag, when going shopping. Love this book. Great for families that have children with allergies.
Lorna

I handed this over to my daughter-in-law. We look at our grandson and then decided to "educate" her. She is still shocked over the crap, no the poisons that are in everything including vitamins.
Peter Bratko

Life Saver!!!!!!! I've bought two copies of this book, one to carry with me every time I go shopping, the other to stay at home whenever my family needs it It's an excellent book, the perfect little gift for someone you love and really care about.
Laura

Ya gotta be kiddin'. For starters, when I use this book as a reference, 100% of the "All-natural", 'Organic' foods that claim "NO MSG" did in fact have it! ... There are many sources to find out the many forms of MSG, and this book lists those that I've found, and why it can occur naturally through various processing methods. Absolutely amazing.

The guide fits in a back pocket or purse and is an invaluable resource. I would even suggest having one in your kitchen and a second copy in your car (or scooter). I'm sure some other source SOMEWHERE has similar information, but I haven't found a resource that cross references itself in a basic format, with terse language (exactly what you need when shopping).

The sad part about this though is that I ONLY purchase 'whole', 'organic', 'healthier' snacks, and ALL the chips and cookies fail her additives test, quite poorly actually. Lesson learned: If I want high-quality goodies without fillers, seems like I'm gonna have to make 'em myself!!!
W. Jackson

"I recommend everyone carry this with them in their purse or glove compartment for handy reference."
The late Lendon H. Smith, M.D.
Author of Feed Your Kids Right

*"**A great pocket resource in understanding labeleze.**"*
Earl Mindell, R.Ph., Ph.D.
Author of The Vitamin Bible

This little book should be in every food buyer's possession. I highly recommend it. Anyone who takes the responsibility of buying food ... should have this book as a guide. ***It gives the bottom line of what many of the common chemicals that are used in the processing of our foods do to our bodies***. You don't have to be a PhD to understand what's spelled out so well in this pocket size book, you only have to have the courage to follow its inherent message and be well on your way to being healthy again."
Razel

"Even though you know instinctively that many food additives are bad for you, you will be shocked when you actually read what these things can do to you. ***The rating system is very helpful*** to know which items are poorly tested and unsafe to eat, versus those that may only cause an allergic reaction. The book is small and I carry it with me every day."
A reader from Sydney, Australia

"***This book is fabulous*** and everyone should not only read it but carry it with them when they shop."
Chelsea Thompson

"I loved your little book. ***Its contents are excellent***...."
Bruce West, D.C.
Author of HEALTH ALERT newsletter

"This is a nice compact reference for all the times you are shopping and wonder what those mysterious ingredients are in your foods. The reference ***tells you which ones are safe and which ones will be harmful***... An easy way to get educated while you shop or looking through your food pantry to make sure you consume what is safe."
Lisa Jeanne, Walpole, Ma

"I'm a graduate student in Nutrition and have found your booklet ***most helpful compared to others*** that I've used."
Brenda L. Johnson

"I am very pleased with this book! At first it seemed like a lot of money for a pocket-sized-book, but it is worth every penny, even with the shipping cost added to the purchase. I am learning a lot, and ***it's so easy to follow, and so very helpful***, plus it's easy to carry, just put it in your pocket. I have highlighted the safe additives for quick reference. ***I highly recommend this book to everyone!***"
Meh

This book provides a ***plethora of information*** on a variety of additives ... Very ***small and easy to flip through*** ... Try it ... By the way, I found my copy at Whole Foods.
Merlot